The Unforgotten Sisters

Female Astronomers and Scientists before Caroline Herschel

More information about this series at http://www.springer.com/series/4097

Gabriella Bernardi

The Unforgotten Sisters

Female Astronomers and Scientists before Caroline Herschel

 Springer

Published in association with
Praxis Publishing
Chichester, UK

Gabriella Bernardi
Torino, Italy

SPRINGER PRAXIS BOOKS IN POPULAR ASTRONOMY

Springer Praxis Books
ISBN 978-3-319-26125-6   ISBN 978-3-319-26127-0   (eBook)
DOI 10.1007/978-3-319-26127-0

Library of Congress Control Number: 2016932888

Springer Cham Heidelberg New York Dordrecht London

Cover design: Jim Wilkie

Printed on acid-free paper

Springer International Publishing AG Switzerland is part of Springer Science+Business Media (www.springer.com)

To all curious and persevering minds

Introduction

The Muses Urania and Calliope by Simon Vouet approximately 1634

March 4, 1845.
MY DEAREST NIECE, –

*   *   *   *   *

Have I understood you alright? Saw you the thermometer 1½° above zero? the lowest I have heard of here was only 13° below freezing; but we are buried in snow!

March 5th. – No alteration in the weather, nor in my affection for my dear niece and nephew and their ten children! the first is as cold as the latter is warm!

*   *   *   *   *

April 29, 1845.
In his father's library my nephew must have found a folio volume of H— (an astronomer and copper engraver), where, for every hour a distinct picture [of the

moon] is given. In the Phil. Transactions for 1780, p. 507, is the first paper of William Herschel on the Moon. In 1787; 1792, p. 27; 1793, p. 206, measure of mountains, &c.

Twenty-three years ago, when first I came here, I visited Madame W. (not von) once or twice, saw her observatory and a telescope, I believe not above 24-inch focal length; at that time she amused herself with modelling the heads of the Roman Emperors: her daughter, then a girl, was a poet, and a portrait of her was exhibited as a Sappho crowned with laurels....

These lines are part of the correspondence between a famous astronomer who lived at the turn of the eighteenth and nineteenth centuries, Caroline Herschel and her niece-in-law, Margaret Brodie Stewart, wife of the astronomer John Herschel, son of her brother William.

Margaret Herschel (Margaret Brodie Stewart) (1810–1864) wife of the astronomer John Herschel, by Alfred Edward Chalon 1829

Caroline Lucretia Herschel (1750–1848) was probably the most famous female astronomer of the past. Her brother, William, became famous for the discovery of Uranus, and together they pioneered a study of the "physical sky," dealing with a background of stars that until then had been considered little more than a stage in witnessing the motions of the planets. They initiated the study of the interstellar cloud, those regions apparently devoid of stars that we now know to be huge regions of space filled with dust that hide the objects behind them, and they investigated the distribution of the stars in space, attempting for the first time to reconstruct the shape of our Galaxy.

Artist impression of Caroline Herschel taking notes as her brother William observes on March 13, 1781, the Night William Discovered Uranus

Among Caroline's correspondence, the above letter is surely not the most important one from the astronomical point of view, but quite interestingly it is one of the very few, possibly the only one, where she mentions another female astronomer, although probably just an amateur one. It is known, however, that Caroline Herschel was not the only female astronomer of the past deserving an appropriate recognition. The feeling of such lack of appreciation is well portrayed in an intense poem by Siv Cedering (1939–2007) entitled "Letter from Caroline Herschel (1750–1848)" where the artist imagines that the famous astronomer is writing a letter telling her about the demanding work with their brother, that involved both telescope construction and night observations, and how it was difficult to find some time for her personal researches. The conclusion betrays her resigned fear of being forgotten, like the five "long lost sisters" of the past she cites:

Aganice of Thessaly, Hypatia, Hildegard, Catherina Hevelius and Maria Agnesi, all of them female scientists and astronomers.

It is satisfying to imagine that this letter is addressed to Mary Somerville, another female astronomer and her contemporary, who went down in history with Caroline not only for their contributions to astronomy, but also for being the first women whose scientific merits received a high academic recognition in the form of honorary memberships of the Royal Astronomical Society.

Actually, despite the many difficulties caused by living and working in a male-dominated world, a remarkable number of women scientists who brought important contributions to the development of science can be accounted. The story hands down to us the names of at least twenty famous scientists of antiquity, among which that of Hypatia is probably the best known.

Detail from The School of Athens (1509–1510) by Raffaello Sanzio. It may be a portrait of Francesco Maria della Rovere, or possibly the philosopher Pico della Mirandola. Although with little support, it has also suggested that it might also represent Hypatia

We can find only a dozen of them in the Middle Ages, especially in convents, but almost none between 1400 and 1500. The counting starts to become more precise afterwards, with 16 in the seventeenth century, 24 in the eighteenth century, and 108 in the nineteenth century, while currently about 2000 women are professionally involved in the field of astronomy alone, as reported at the International Astronomical Union General Assembly of 2015. (Actually, among the 11,273 members of this organization 1792 are women, i.e. about 15.9% of the total number.). But how many of them are cited in textbooks?

Indeed, there are no clues of similar acknowledgments to other female figures before Caroline Herschel and Mary Somerville, and unfortunately the history of women in culture, as well as in civilian life, is one of exclusion up to the end of the nineteenth century and still largely up to the middle of the twentieth century. That is at least in industrialized countries, since in many developing nations, with rare exceptions, women are still far from achieving even their most basic rights as human beings. This book will not deal with the causes of this situation, rather it will describe the lives of these distinguished scientists and their achievements in science. Not without difficulty, they undertook mathematical and astronomical studies at a professional level, sometimes leaving an indelible mark in the history of science, despite the obstacles mainly due to the ideas and preconceptions that society placed on them and not on their male counterparts. Indeed, for centuries those women who had access to education were highly placed and lived in monasteries, where actually a fair dowry was required. In other historical periods those few women, favored by having a father, a brother or a husband scientist willing to share their knowledge, could acquire a scientific education, but still in the early twentieth century, in many European countries, girls were denied access to universities and also to high school. Consequently, since women were excluded from universities and from a regular scientific education, there emerged only a few rare cases where they were given a chance, giving rise to the notion that women are not suited to scientific subjects.

The Three Rs

This book tells the life stories of 25 female scientists, mainly astronomers and mathematicians, who made very important contributions to the development of science, yet for too long remained forgotten. For each of these "long lost sisters of science", to use Caroline Herschel's phrase in the poem dedicated to her, the author has organized a form of personal file that places each subject's life within its historical context, outlines her main works, highlights some curious and interesting facts, and presents comments from contemporaries and descendants. The book reaches back more than 4000 years, to En HeduAnna, the Akkadian princess, who was one of the first recognized female astronomers, and includes such luminaries as Hypatia of Alexandra, Hildegard of Bingen, Elisabetha Hevelius, and Maria Gaetana Agnesi, through to Mary Somerville and Caroline Herschel herself. The book will be of interest to all who wish to learn more about the women from antiquity to the nineteenth century who played such key roles in the history of astronomy and science despite living and working in largely male-dominated worlds.

Approaching scientific disciplines, and in particularly those related to astronomy, cannot be attempted without some basic knowledge. In this sense the first two of the "three Rs," that is how to read and write, is not enough, and the third one has the most prominent role. This has to be intended in a sense broader than the original one. Indeed, one has to juggle in a mathematical universe made of arithmetic, algebra, geometry, angles and equations of all kinds. Without a proper education it is impossible to move up in this world, so being excluded from knowledge's distribution channel meant to be cut off from certain areas. For centuries, the transmission of knowledge happened through circuits which were quite different from those we currently consider the normality. There was nothing like a minimum level guaranteed to everybody by some public institution. Education, until the nineteenth century, was basically a private burden that could be supported only by families that had both economical means and thirst for knowledge. Education was therefore a privilege usually reserved to the higher social classes. In addition, the kind of education one could receive, and thus win eventual access to the scientific community, depended on the cultural preparation and the social position occupied in the civil community. This situation would be enough to understand why so few women are remembered in the history of science.

To give an example, for a long time women were denied access to any University, and an opening occurred for the first time in Europe in 1867 at the Ecole Polytechnique of Zurich. Before then, only Italian Universities had awarded academic titles to some women considered as special, such as the Venetian noblewoman Elena Lucrezia Cornaro Piscopia, who was the first woman in the world to get a master's degree in Philosophy from the University of Padua in 1678.

Portrait of Elena Lucrezia Cornaro Piscopia, first women graduate in the world

Despite these difficulties, there have always existed women scholars who worked in scientific disciplines, if they were just given a minimum of freedom and power to access the necessary studies. For centuries, however, they were mere exceptions, and moreover their names have been erased from history. In almost all circumstances they were daughters, wives or sisters of scientists, and their contributions were often confused with their male relatives.

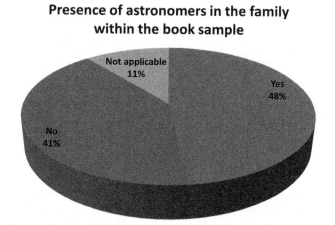

Presence of astronomers in the family within the book sample

In some cases, where it was possible, they also devised methods to be taken into consideration by publishing under a pseudonym. The most famous example is that of the mathematician Sophie Germain who, during the nineteenth century, signed herself as "Monsieur Le Blanc" in order to communicate with the community of mathematicians such as the celebrated Louis Lagrange.

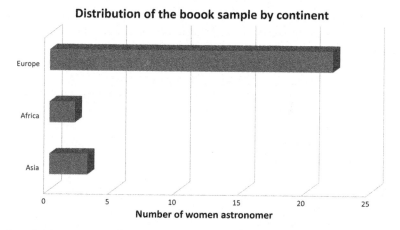

Distribution of the boook sample by continent

The aim of this research was to describe the lives and the works of the most notable astronomers of the past. They lived mainly in the European-Mediterranean basin, from antiquity to the eighteenth century, until a period ideally represented by Caroline Herschel at a time in which the idea of a more public role of women in Astronomy started to take its first steps.

**Distribution of the book sample
by historical period**

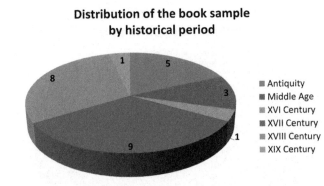

These women undertook their mathematical and astronomical studies at a professional level, distinguishing themselves, not without difficulty, often leaving important contributions to the history of science, despite the many obstacles created by the ideas and preconceptions that society burdened them with in contrast to the advantages of their fellow men.

It was interesting to learn more about the family relationships of some of these scientists and to analyze their influence on the personalities and scientific contributions of these researchers. Often a link was created by some mutual interest between father and daughter, husband and wife or brother and sister, and these strong emotional ties with a prominent but friendly male figure affected the life and the scientific interests of these women who would otherwise not have been able to develop on their own.

We are quite used to this kind of separation of roles between men and women that, among other avocations, reserved the study of natural sciences to the male part of society. This ethos can be dated back to such ancient times that, to modern eyes, female involvement in these subjects appears to be a novelty of our own times. It is known, however, that in antiquity there existed matriarchal societies in which female deities were worshipped in the belief that they were responsible for creation. These societies gradually disappeared, at least in the Mediterranean basin, in favor of prevailing cultures in which religion attributed the creation of life and the universe to an exclusively male principle. It might then be that the study of nature was different in the past, and that women could have taken part in it even as it began a natural removal of themselves from these activities. It might also be that the ancient priestesses of the Mesopotamian and Egyptian civilizations, of which some examples are cited in this book, can be interpreted as the last traces of these ancient roles, which went preserved for some time thanks to some kind of religious legacy.

Such figures probably became less and less common with the evolution of more "modern" societies, but despite the increasing obstacles women had surely to meet, in particular for their participation in scientific knowledge, female contributions never came to an end. Even in the strongly misogynist Greek society we can find some of them back in the seventh and sixth century BC. They flourished within the Pythagorean school, which admitted also women as disciples, among whom the most famous name is that of Teano. Philosopher, mathematician, astronomer and physician, she is credited with a prominent role in this school and of several works in these disciplines, but the available information is extremely uncertain because of the sectarian character of this organization and of the secrecy imposed on its followers. Not even her degree of kinship with Pythagoras can be clearly established: she could have been his wife, but in other sources she is attested as his daughter. In any case, such a relationship allowed her to study and to became a scientist who could succeed Pythagoras himself in the direction of his sect in the sixth century BC.

Among the other figures analyzed in this essay, one of the most famous of antiquity is Hypatia of Alexandria, who lived between the fourth and the fifth centuries AD. She was a mathematician, astronomer and philosopher who was brutally killed as a consequence of the fight for power between the pagan Hellenic faction and the emerging Christian sect which characterized that historical period. She is mentioned together with Aganice of Thessaly, Hildegard of Bingen, Catherina Hevelius and Maria Agnesi in the poem "Letter from Caroline Herschel" by Siv Cedering cited in the introduction, but as we can see from our list this represents just a limited sample.

In the early Middle Ages the social position of women within noble families remarkably improved, at least in certain aspects. The monasteries, especially, for a certain period represented remarkable cultural centers and environments where women in certain cases could be educated in every field. The most significant example of such a situation is Hildegard of Bingen. In the following, due to a setback in the sexist sense of the Christian Church, the female role was increasingly hampered and consequently became more and more marginal, so that when, at the beginning of the eleventh century, the first universities were established, women were immediately excluded and had to resort to forms of non-academic education. It is especially in this way that there developed the above mentioned form of partnership between men and women in which the former played the dominant role, with mostly theoretical knowledge, and the latter instead became their assistants with predominantly practical or secondary tasks.

In the sixteenth century new and revolutionary concepts were developed in astronomy, and also several female scientists took part in this enterprise, although at the same time "witch hunts" were storming in all Europe. Later, the seventeen and eighteenth centuries witnessed the appearance of a new phenomenon among the women of aristocracy, that of scientific circles or "salons" held by women for scientists and scholars.

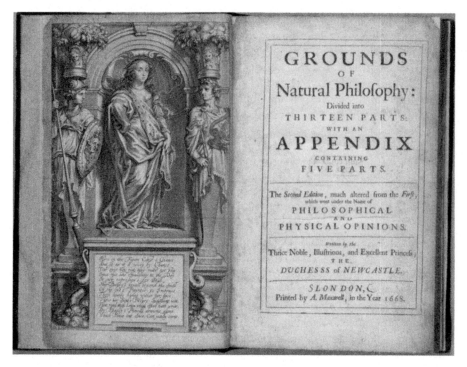

Title page and frontispiece from "Grounds of natural philosophy" by Margaret Cavendish, Duchess of Newcastle

Among the "dames of science" many prominent figures can be found, such as Margaret Cavendish (who financed the Royal Society and the Lucasian Chair of mathematics at Cambridge University, the place held by Newton), Lady Conway, Queen Christina of Sweden or the marquise du Châtelet.

Queen Christina of Sweden (*left*) and René Descartes (*right*)

All of them distinguished themselves not only in the role of patroness, but also in that of scholars. Quite ironically, this phenomenon also contributed to the birth and development of the great scientific academies and societies, which, however, (at least the most prestigious ones) did not admit the presence of women until the twentieth century.

Opportunities were developed in relevant scientific research that emerged where innovative methodologies and techniques, especially in the world of crafts where women's skills could be exploited. Examples of this phenomenon include Elisabetha Koopman Hevelius and Elisabeth Maria Winkelmann Kirch, during the seventeenth century. The development of science made the equipment more and more complex, and when its costs exceeded those accessible to a personal and "homemade" or craft enterprise, the research shifted permanently within the Academic and male-dominated world. Similarly to what happened with the birth of the Universities or with the scientific societies, women were cut off from the official and most relevant scientific activity, and once more became reduced as ghost aides. Another reason that explains why several of them are unknown nowadays.

The modern era witnessed significant social changes, like emergence of a bourgeois social class and its values, which became prominent in society together with the importance of this new caste. However, their social canons relegated the women to a role of family management, and as a kind of "social showcase" whose prestige essentially depended on that of the man, to whom was exclusively reserved the leading and working roles. Women's education was not necessarily included in these social obligations.

Acknowledgments

I want to thank Dr. Alberto Vecchiato from Astrophysical Observatory of Turin for giving us both his time and the benefit of his scientific, historical and languages knowledge. I am also grateful for the help of many librarians, archivists and their Italian and foreign institutions.

Contents

Part I
Timelines from Enheduanna to Sonduk

2400 BC introduction of sexagesimal numbering

2000 BC Stonehenge circle

1800 BC Code of Hammurabi

1600 BC Shang dynasty begins in China

1350 BC Phoenicians introduce an alphabet

1200 BC beginning of the iron age in the Middle East, The Trojan war

776 BC first Olympiad

540 BC Pythagorean school was born in Crotone

538 BC the Persian conquered Babylon

480 BC Parmenides theorises a spherical Earth, Battle of Thermopylae

400 BC the end of the Olmec culture in Central Mexico

332 BC foundation of the city of Alexandria in Egypt

264 BC first Punic War begins

285 BC death of the Greek mathematician Euclid

215 BC beginning of construction of the Great Wall in China

200 BC Eratosthenes estimates the Earth's circumference; invention of the armillary sphere

150 BC Hipparchus publishes his star catalogue and discovers the precession of the equinoxes

44 BC Julius Caesar's death

150 AD Ptolemy publishes the *Almagest*, the Geocentric theory

450 AD founding of the city of Teotihuacan in Central America

476 AD fall of the Roman Empire

Chapter 1
Enheduanna (XXIV BC)

It's time to rekindle the stars (Guillaime Apollinaire-Les mamelles de Tirésias)

We will never be able to know for certain the identity of the first woman who, hidden by the mists of time, worked systematically in the field of astronomy. However, the name of a Mesopotamian Princess has emerged from ancient clay tablets. Enheduanna or Enheduana, En-hedu-ana, En HeduAnna; this is the name of the woman bearing the first documented witness of a female astronomer in antiquity. We don't know much about her, but she lived around 2300 BC in the Sumerian region and probably her father was a king, Sargon I, the Great (2335–2279 BC). He came from the city of Akkad and was the founder of a dynasty which, about 4000 years ago, unified different people of the Mesopotamian region into a vast empire, including all the Sumerian cities. Sargon was not a noble, but rather a social climber. Starting as cupbearer of the King of Kish, a Sumerian city, he eventually became the *Lord of the four kingdoms*, creating an empire that stretched from the Persian Gulf to the Mediterranean including Mesopotamia, Elam, Syria, Oman, Phoenicia and part of Anatolia to the river Hylas. After being proclaimed God, son of Inanna, the Sumerian Goddess of love and fertility whose Babylonian counterpart was Ishtar, he ruled his empire in a balanced way, respectful of the traditions and customs of the different people subjected to his authority. He had the foresight to surround himself with men of great merit and although he was a man of war, he supported art, culture and science. The monarch appointed his eldest daughter, Enheduanna, *High Priestess of the Moon goddess in the City*, a position of considerable prestige. Indeed we must remember that priests and priestesses played a fundamental role in the Mesopotamian civilizations. From their sacred temples they were not only the depositories of knowledge, but also directed all-important activities such as trade, agriculture and handicrafts. It is because of this role that Enheduanna can be regarded as an Astronomer, one of the activities falling under her authority.

© Springer International Publishing Switzerland 2016
G. Bernardi, *The Unforgotten Sisters*, Springer Praxis Books in Popular Astronomy,
DOI 10.1007/978-3-319-26127-0_1

1.1 Babylonian Astronomy

The designation of "Babylonian astronomy" is conventionally adopted to identify
this science and its development over a long period of time, beyond that of the
civilization bearing the same name, extending over more than 3000 years from the
Sumerian period to its last legacy belonging to the Seleucid era, thus mixing up with
and influencing the achievements of Hellenistic civilization.

Babylonian tablet at the British Museum in London, recording Halley's comet during an appearance in 164 BC

The "Babylonian" attribute characterizing this definition is given to highlight the
outburst of discoveries and systematization of the knowledge which grew during

their era. Among others, to Babylonians must be ascribed the merit of having created a network of observatories in which priests and priestesses studied the movements of celestial bodies, possibly creating the earliest star catalogue and the first examples of celestial maps. Their observations also allowed precise measurements of several celestial events such as the phases of the Moon and the lunar month, the Solar year and planetary motion. The former were at the basis of their definition of the month, whose starting moment was marked by the first day after the new Moon, when the first Crescent Moon Scythe appeared after sunset. Altogether, they discovered some important cycles used to predict the eclipses and lying at the basis of their calendar. This is of luni-solar type, based on a 19 years long cycle called "Metonic" after the name of a Greek astronomer who actually learned it from the Babylonians, and it allows the definition of a calendar that can keep the lunar months and the seasons aligned. The same algorithms used in this calendar are still used to set the dates of certain religious events such as the Christian Easter and Passover. The Babylonians also introduced the way of subdividing the day we still use, with 24 h, each composed of 60 min, which in their turn were divided into 60 s each. This subdivision is linked to the way they represented the numbers which adopted, for the first time in history, a positional notation with a sexagesimal system, similar to our decimal one, but whose base is 60. The great advantage of such a notation is that not only does it permit the representation of large numbers with a limited set of symbols, but also greatly eased mathematical calculations. The Babylonians also established measures of length, of weight that, together with the time already mentioned, were the basis of the knowledge of other cultures.

In a later period, around the fifth century BC. they noted that the apparent motion of the Sun and Moon, from West to East, happened at a variable rather than at a constant speed: the two celestial bodies seemed to accelerate in the first half of the apparent motion of revolution, up to a maximum value, and to decelerate in the remaining half, eventually regaining their initial velocity. In order to explain this phenomenon, Babylonian astronomers formulated the first mathematical models on the motion of the stars, which were used to predict the time of a new moon and the start of each month. Thanks to their long tradition in the study of the positions of the planets, the Babylonians noted that these celestial bodies periodically came back to the same place with respect to the fixed stars, and named these periods "great cycles," determining those of Mars, Mercury, Jupiter, Saturn and Venus.

The main source of information about Babylonian astronomy are hundreds of clay tablets written with cuneiform characters found by archeologists on their excavations. They record both the astronomical observations and the calculations made by these ancient priests and scientists. In particular, about 300 tables, mostly from Uruk and Babylon, were discovered in the nineteenth century and sold to the British Museum. It is on such findings that the planetary theories could be unveiled. Moreover, they show that both the astronomers of Uruk and those of Babylon were very active during the late Seleucid period, that is the last of the Babylonian astronomy, when this ancient science merged with Greek culture, bringing the achievements of the Hellenistic science. The former in fact can be dated between 200 and 160 BC, while most of the Babylon tablets belong to the period between

170 and 50 BC. Gaius Plinius Secundus, better known as Pliny the Elder, was a Roman author, naturalist and a natural philosopher who died during the eruption of Vesuvius in 79 BC. He wrote an encyclopedic work, "*Naturalis Historia*", Natural History, says that there were three schools of Babylonian Astronomy: Babylon, Uruk and Sippar, but so far there has been no evidence of astronomical discoveries of this new center.

1.2   Works

Unfortunately very little is known about the life and the astronomical knowledge of Priestess Enheduanna. Appointed by her father as *High Priestess of the Moon Goddess of the City*, she devoted herself to the study of movements of the Moon and the stars, and together with other priests of the Akkadian Empire, she established an extended network of observatories which monitored systematically the movement of the stars. We know that the observations and the algorithms which allowed construction of the Babylonian calendar can be dated back, at least in part, to the period of Enheduanna. Their origin can thus be attributed to this ancient community of priests and priestesses who observed the sky from their temples, however we cannot single out the exact contributions of our first known female astronomer, and this is not only because she lived such a long time ago. As a matter of fact we cannot identify specific contributions of *any* single Mesopotamian astronomer because of the communitarian character of these societies. For this reason everything was centered on the king and on the prevailing religion, and the achievements and discoveries were used without any personal attribution. Such was the state of things at least until the advent of the Seleucid era, when we start having some exceptions, among which we can find the names of the only two Babylonian astronomers cited for their contributions. This therefore explains why, unfortunately, we do not have any scientific writing of Enheduanna, but only 48 poems or hymns. One of these poems, however, is more than sufficient to witness the importance of this woman in her society and her specific activity in the field of Astronomy:

> The true woman who possesses exceeding wisdom,
>   She consults a tablet of lapis lazuli
>   She gives advice to all lands...
>   She measures off the heavens,
>   She places the measuring-cords on the earth.

1.3 Curious Facts

Enheduanna is indeed just one name of a presumably long list of women who studied the stars and the Moon cycles, even if hers unfortunately is the only one which survived to the present days. This almost certain fact is due to the relatively large cultural freedom allowed to women in the ancient Mesopotamian societies. Actually, among the Sumerians, women could reach quite a high social position and they were granted a good autonomy, and from the Hammurabi Code we know that they could carry on commercial activities, own property and become judges and sages.

Enheduanna is portrayed in a bas-relief in Philadelphia at the University Museum showing her while presiding over a religious ceremony. This alabaster disk depicts the priestess in the act of celebration of a religious ceremony in honor of the goddess Inanna, confirming her pivotal role within the community and in the temple of Ur, one of the most important religious centers of Sumer. This disc is in fact the symbol of Nanna, the Sumerian goddess of the Moon which, according to one tradition, was the mother of Inanna, and it was found by Leonard Wolley in 1925, in the architectural complex of the priestess of Nanna at Ur. Enheduanna resided there as the Sun Priestess of that Goddess, to whom her life had been consecrated. She is portrayed in the traditional clothing of her role with two servants near her, while offering ritual libations to the goddess Inanna. On its back the disk has an inscription where the Priestess claims to be the daughter of Sargon, the King of all and the real Lady of Nanna.

1.4 As They Said of Her

The meaning of the name of Enheduanna, is "en" means High priest or high priestess, "hedu" means adornment, so this name translates to *"high priestess adornment of the god An"*. In particular "An" was the sky god.

In antiquity, power, religion and science were tightly linked, and the very prestigious position of *High Priestess of the Moon Goddess of the City*, made her one of the most powerful persons of the Sumerian cities. Her power can be well understood by noticing that only through the auspices of the High Priestess was a king entitled to rule, although in the case of Enheduanna it was her father.

Chapter 2
Aganice (XX BC)

> *Do you not know, Asclepius, that Egypt is an image of heaven
> or, to be more precise, that everything governed and moved in
> heaven came down to Egypt and was transferred there? If
> truth were told, our land is the temple of the whole world.
> (Hermes Trismegistus-Corpus Hermeticum)*

Aganice, cited in other texts as Athyrta, is the name of an Egyptian princess who
lived around 1900 BC, during the Middle Kingdom (about 2000–1700 BC) working
on astronomy and natural philosophy. As it is reported in ancient religious writings,
she is supposed to be the daughter or the sister of the king Sesostris I or Sesostre of
Twelfth Dynasty. This Pharaoh ruled Egypt for 45 years, pursuing an expansionistic
policy through several military expeditions that enlarged and secured the borders of
his kingdom. He was the first Pharaoh to leave the capital city to go to the lands of
Kush, in Nubia, but he never managed to conquer them. Despite the military
activity at the borders, the internal policy was characterized by a peaceful climate
which favored the development of arts and science, thus continuing a long-standing
tradition whose origins can be traced back to the Old Kingdom, which lasted for
about 500 years starting from the twenty-seventh century BC.

© Springer International Publishing Switzerland 2016   9
G. Bernardi, *The Unforgotten Sisters*, Springer Praxis Books in Popular Astronomy,
DOI 10.1007/978-3-319-26127-0_2

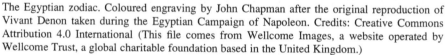

The Egyptian zodiac. Coloured engraving by John Chapman after the original reproduction of Vivant Denon taken during the Egyptian Campaign of Napoleon. Credits: Creative Commons Attribution 4.0 International (This file comes from Wellcome Images, a website operated by Wellcome Trust, a global charitable foundation based in the United Kingdom.)

Women were an integral part of this society, to such an extent that there were medical schools in which they could be educated. There are distinguished examples of this institution, including Moses and his wife Zipporah, who are claimed to have studied medicine at Heliopolis around 1500 BC, and Hashepsut, woman Pharaoh of the Eighteenth dynasty. Indeed, she was called the Queen-doctor, and promoted a botanical expedition searching for officinal plants. In Egypt, not just medicine but also observation of the stars had very remote origins. In the southern Sahara desert, near Nabta in the Nubian desert, what is believed to be the oldest Astronomical Observatory, erected well before the Age of Pyramids, is still standing after 7000 years. The site consists of a small stone circle, with a series of grave-like flat structures, and of five lines of megaliths that resemble those of Stonehenge and of other European areas which, however are dated 1000 years later.

2.1 Astronomy and Mathematics in Ancient Egypt

The proto-Egyptian site of Nabta, which appear connected to the Solstices, may witness a kind of astronomy with an important religious significance. Although the connection with religion, as usual for ancient people, was very tight for Egyptians as well, they nonetheless showed a very practical approach with regard to this discipline, given by the need of exploiting the observation of celestial bodies to solve the agricultural problem par excellence, i.e. the period of recurring floods of the Nile. Available documents suggest that in ancient Egypt the heliacal rise of Sirius (the term "heliacal rise" means that an object in the sky rises in the morning, just before Sun) was regularly observed and recorded, and it represented an integral part of the Egyptian agricultural calendar because this event heralded the flooding of the Nile, an event with utmost importance for the agricultural economy of the people.

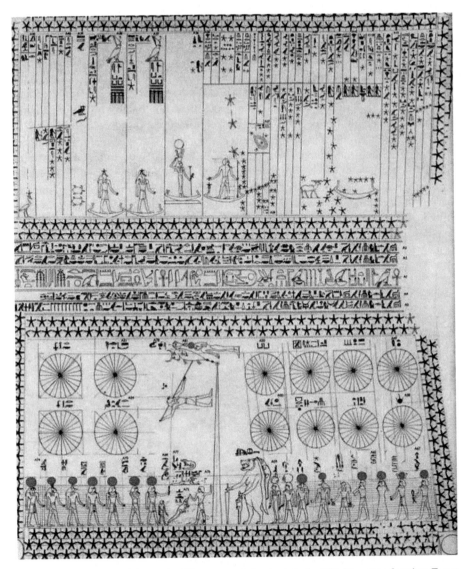

The double wall mural of the tomb of Senenmut, dating back to the 18th Dynasty of ancient Egypt (ca. 1473 BC) and depicting circumpolar constellations in the form of discs. Some of the main figures and stars seen in the diagram are Sirius, Orion, Ursa Major, Draco

Unfortunately there aren't many papyri showing their observational methods, but an important number of documents with mathematical content, which might explain why Aristotle, in his *Metaphysics*, wrote that the Egyptians were the greatest mathematicians in history. One of the most important examples in this field is given by the Rhind Papyrus (bought by a Scottish lawyer in 1858 in Luxor) written by the scribe Ahmes, who was a mathematician under the Hyksos Kingdom of Apophis, around 1550 BC. from a now lost document dated about 400 years back.

Written most likely with didactic purposes, it is structured as a sort of manual divided into five sections: arithmetic, stereometry, geometry, calculation of the pyramids and a set of resolutions of practical problems for a combination of 87 problems. Another mathematical document, the Moscow mathematical Papyrus (ca. 1850 BC) poses a highly complex issue, that of calculating the volume of the truncated pyramid, which will happen only at the time of Euclid 1550 years later. Egyptian astronomy, like that of the Babylonians, covers a very long epoch which stretches at least from the Old Kingdom until the first centuries of the common era, thus including the Hellenistic period of about five centuries when, thanks to the strength and wisdom of the Greek conquerors, the knowledge of these three cultures merged and further influenced each other. For example, starting from 300 BC the first Babylonian-Egyptian zodiacs appear on the temples' ceilings, the most famous of which is that of Dendera.

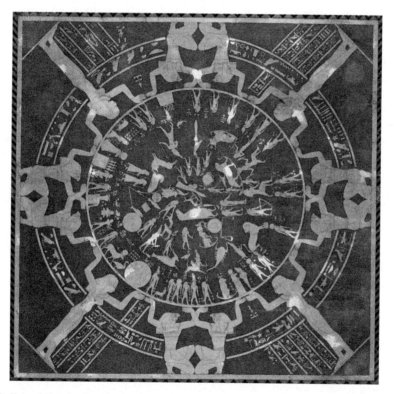

The Zodiac of Denderah reproduced on one of the ceilings of the mythological Hall of the Egyptian Museum (Neues Museum) at Berlin. Credits: Creative Commons Attribution 2.0 Generic (Jean-Pierre Dalbéra)

Also, documents written in Greek and Demotic (the cursive hieroglyphs) of astronomical-astrological topics, can be dated around 200 BC, and another papyrus dated 144 AD shows the phases of the Moon.

2.2   Works

Little is known about the astronomical activity of Aganice, with the exception of her attempts to predict the future using celestial globes and of her studies of the constellations, however this has to be set in the framework of how science was intended in this civilization. The Egyptians were interested primarily in the practical aspects of science. Priests and priestesses of sacred temples used mathematics and astronomy mainly to address concrete issues, such as definition of the agricultural calendar, irrigation, wheat storage, or orientation of obelisks and the cut of stone blocks used for their impressive temples and monuments. Unfortunately no documents, such as the Rhind Papyrus for mathematics, have been found about their astronomical knowledge. What we know of this subject comes from the so-called "Book of Nut," a treatise on religious astronomy containing also a description of astronomical observations, and from paintings and inscriptions found in several tombs and sarcophagi. It is from this material that, for example, we know of the "diagonal star clocks" time-keeping method which made use of star ephemeris which appeared for the first time in the Old Kingdom. Studies on the orientation of pyramids and the development of instruments such as the hourglass and water clock, the merkhet (see Curious facts) and sundials helps to cast light on various aspects of the Egyptian astronomical knowledge, but observational evidence always appears in pictorial form. It has to be highlighted, however, that the way science was intended in ancient times was quite different from now, and science and religion were tightly connected. The practical inclination of Egyptians therefore encompassed also religious traits because the prediction of the future (what we would now call astrology) was not so different from the prediction of the Nile floods. They were both intended as an interpretation of the Gods' wills. The activity of Aganice can therefore be perfectly understood within this more ample sense. Moreover, one has to take into account that Egyptian astronomy was quite influenced by that of the Babylonians, to whom can be ascribed many important astronomical discoveries, as we have seen in the chapter about En-Heduanna, but also the "invention" of the Zodiac and of astrology in the sense it is intended today.

2.3   Curious Facts

According to some sources, the main astronomical instrument of the Egyptians was the merkhet. Its origin is quite remote, dating back to 2600 BC and its simple design consisted in a palm leaf cut on the top and a set square with a plumb line. It was used to fix the orientation of monuments such as temples and pyramids, to observe the meridian transit of the stars, and for the measurement of fields. It was also used to determine night-time. Two or more observers were seated at a fair distance, one in front of the other, oriented along the north–south direction, and holding the instrument in his hands, where the palm tree rib served as a viewfinder which sighted the stars that culminated through the plumb line of the square. Referring to the shape of the observers who turned their backs on the South, an aide read the hour according to the position that the star had on the table. It is most likely thanks to this tool that the pyramids could be aligned with a very high degree of accuracy with cardinal points. The pyramid of Khufu, for example, has an error of just 3 min of arc!

2.4   As They Said of Her

In a poem by Siv Cedering, the famous astronomer Caroline Herschel is imagined remembering Aganice as one of her forgotten sisters from the past. However, since she is cited for having been accused of witchery, it is more likely that the author is referring to Aglaonike of Thessaly instead, as we will see later.

Chapter 3
Theano (Sixth Century BC)

Je vous adore, ô ma chère Uranie!
Pourquoi si tard m'avez-vous enflammé? (Voltaire, Épîtres,
stances et odes)

Despite the obstacles women had to undergo in order to access the scientific knowledge in a strongly misogynist Greek society, there are evidences of female contributions which date back to seventh to sixth century BC. They come from the Pythagorean school, which represents a very particular case among the educational institutions of this civilization, also because both sexes were admitted among their members and among the women the most famous is Theano, even if several other female memberships of this school are known.

We are in the so-called "Magna Graecia" (which in Latin means "Great Greece") and precisely in the current Italian city of Crotone, Calabria. Here Theano was born around 550 BC, when the city still had its original name of Kroton. She was a member of the Pythagorean school, which was founded in her hometown in 525 BC, but although it is well known that Theano became an important philosopher, mathematician, astronomer and physician of this school, other information about her is fragmentary.

© Springer International Publishing Switzerland 2016
G. Bernardi, *The Unforgotten Sisters*, Springer Praxis Books in Popular Astronomy,
DOI 10.1007/978-3-319-26127-0_3

Pythagoreans celebrate sunrise, by Fyodor Bronnikov (1869)

Even her family records are uncertain, in fact some sources report her as Pythagoras' wife or daughter, while others claim that she was the daughter of Brontinus, a remarkable and representative man of the city's aristocracy.

3.1　The Pythagorean School and Its Astronomy

Pythagoras, Greek priest and scientist, lived in the sixth century BC and was one of the earliest and most vigorous supporters of the mathematical nature of cosmic order. Practicing music in addition to the study of nature, he began to wonder whether, like that of sounds, the harmony of natural laws could be based on numbers. Such an analogy was based on the observation that sounds produced by stringed instruments are perceived as harmonious when the lengths of the strings are simpler ratios between integer pairs: 1 and 2 for the eighth, 2 and 3 for the fifth, 3 and 4 for the fourth. The numbers could then belong to a world beyond perception, which could be read only with a thought. His bold conclusion was that the first numbers 1, 2, 3 and 4 were the source of all known objects, corresponding to the point (number 1), the line (2), the equilateral triangle (3), and the tetrahedron (4) i.e. the pyramid with four equilateral triangles as faces respectively. According to some sources Pythagoras was also credited for introduction of the five "Pythagorean" solids or polyhedra, later called "platonic solids" named after the famous Athenian philosopher. These include, in addition to the already mentioned tetrahedron, the cube, the octahedron, consisting of eight equilateral triangles, the dodecahedron, formed by 12 pentagons, and the icosahedron, which has twenty equilateral triangles. Represented as points, the number 1, 2, 3 and 4 form an equilateral triangle with four rows of dots called the Pythagorean tetraktys, from tetras, i.e. four in

Greek. For Pythagoras, these numerical and geometrical ratios represented the rational essence of a world composed of four elements, Earth, Water, Air and Fire, punctuated by four seasons, oriented according to the four compass points and crossed by four rivers: the heavenly Pihon, Gihon, the Tigris and the Euphrates. The followers of Pythagoras swore allegiance to "one who entrusted to our soul the tetractys, the source and the original root of eternal nature". The sum of integers of the tetraktys is 10, the number that the Pythagoreans considered perfect. When ten was reached, they went back to 1, the number of creation. These to Pythagoras were the frame numbers of the universe, existing independently of us in reality but also an intellectual key that allows us to perceive and understand the universal harmony.

It is said that Pythagoras had consulted the Oracle of the God Apollo at Delphi who had predestined the city of Crotone as the site of his school, and that was born by the will of that god. The city of Crotone appeared fit for his purposes, as it had developed a prominent scientific and medical culture, and here Pythagoras managed to gain the favor of the people because of his knowledge. The school could also be attended by women on an equal-opportunity basis, and offered two types of lessons: public and private. In public lessons, followed by common people, the teacher explained his philosophy in the simplest possible way, so that everybody could understand its number-based principles, while private lessons were given at the highest level, and they were attended by members already familiar with the more complex mathematical subjects. In this respect Crotone's school, founded by Pythagoras, inherited a mysterious character, but also developed a broad interest in philosophy, mathematics, astronomy and music.

As regards astronomy, the Pythagoreans placed, in the center of the Universe, an immense fire called Hestia. This has an obvious similarity with the Sun, which however was depicted as a huge lens that reflects that fire, lighting up all the other planets. The Sun and all the other planets orbit in circles around this central fire. The first object in order of distance is the Counter-Earth, then the Earth, which is not at rest at the center of the Universe, but just a planet among the others, then the Sun, the Moon, the five planets (Mercury, Venus, Mars, Jupiter, Saturn) and finally the heaven of the fixed stars. The idea of the existence of the Counter-Earth was probably born with the need to explain the eclipse.

Moving according to precise mathematical ratios, the planets generate a sublime and sophisticated sound. Human beings feel these celestial harmonies, but they cannot hear them clearly because they have been immersed in such sounds since their birth. The school had a deep reverence towards the sphere, which was regarded as the material representation of harmony since all its points are equidistant from the center.

3.2   Works

Several contributions of scientific character are ascribed to Theano, but unfortunately no precise attribution can be made in this respect. This is in part due to a somewhat sectarian conduct of the school, which imposed secrecy to its members on specific subjects. We have already seen this character in action when we underlined the uncertainties on the degree of relationship of Theano to Pythagoras himself. Another reason is that all the writings of the disciples were signed by Pythagoras himself, which therefore were often attributed to the founder.

Despite these difficulties, seven letters written by this female philosopher are considered authentic. Their content does not concern cosmology or medicine. Rather, they are a sort of behavior handbook for young brides. This is quite interesting since it can still be considered within the framework of the Pythagorean school. Indeed, women were held to be different from men, although from what we have seen above such differences do not necessarily have to be interpreted as a sign of inferiority. As a consequence, women were taught practical domestic skills in addition to philosophy, which explains why these letters can be interpreted in the context of the school. The advice given here is of different characters: from appropriate behavior towards children to that within a married couple, or from the right way in administration of the family home to the managing of servants.

Among the contributions attributed to Theano, despite the above described difficulties, some are titled "Cosmology" and "Construction of the Universe". These works describe a Universe built through numbers and simple proportions, according to the precepts of the Pythagorean philosophy, and we can find here the astronomical and cosmological model described in the previous section. Finally, while it is quite likely that Theano had produced other treaties about mathematics, cosmology, physics or medicine, no trace of them has survived.

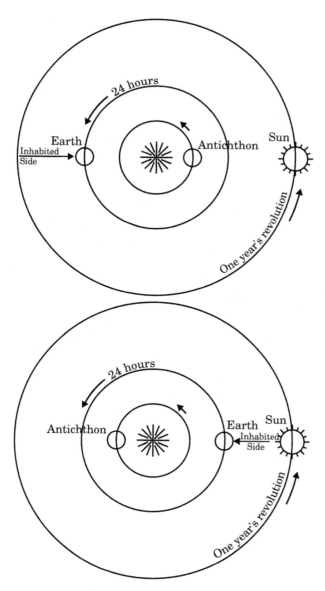

The Pythagorean astronomical system

3.3   Curious Facts

Despite the unclear link with Pythagoras, it is undoubtable that Theano became a distinguished scientist and a prominent member of the School, and some sources report that she succeeded him, on his death, in the direction of this sect in the fifth century BC. She had several female disciples, among whom her daughters Damo and Arignote, who helped her to spread the Pythagorean philosophical and religious system to other women and countries, travelled also in Greece and Egypt. Once again, however, the relationships among these women are often confused. Some of them, like Fintis, can be certainly identified as one of her pupils, but for example the name Mia (or Myia) is remembered as a daughter of Pythagoras even if she was probably just one of her disciples. Also Arignote, according to some sources, was not an actual daughter, and the same fate is attributed to another of her alleged daughters Melissa, who wrote about women's rights.

3.4   As They Say of Her

Teano is considered the most famous cosmologist of the Pythagorean school, as well as an excellent healer. Indeed, in the medical field she is attributed some concepts that will be taken up by physiology in later centuries. An example can be found in a theory of the human body as a microcosm copy of the macrocosm constituted by the entire Universe, a concept which will be taken up during the Middle Ages by the Abbess Ildelgarda of Bingen, who will appear later in this book.

Chapter 4
Aglaonike (V BC or 200 BC)

Yes, as the Moon obeys Aglaonike (Mocking Greek proverb)

Its name means "shining victory" or "victory of light", but other than that very little is known about Aglaonike, also called Aglonice, who was born and lived in Thessaly. According to some sources she is considered to be the first woman astronomer and in fact Aglaonike was a master in the art of predicting eclipses, being able to determine the exact time and locations of lunar eclipses. In this regard she was apparently aware of the lunar cycle of about 18 years, called Saros, discovered by ancient Babylonian astronomers.

Despite Plato's statements in favor of women's education, women did not enjoy the same rights as men. Quite the contrary, Greek society was strongly men-driven, but nonetheless Aglaonike's father did not hamper the interests of her daughter, allowing her to study Babylonian astronomy that had accurately described lunar cycles. With this knowledge she was able to successfully predict eclipses of Sun and Moon defining exactly their times and places.

© Springer International Publishing Switzerland 2016   23
G. Bernardi, *The Unforgotten Sisters*, Springer Praxis Books in Popular Astronomy,
DOI 10.1007/978-3-319-26127-0_4

A "Solar Eclipse" taken by the Apollo 12 astronauts

Her contemporaries were impressed by her predictions, and being unable to understand her knowledge, they considered her as a witch, while in other cases her abilities and the fact of being a woman were the cause of great hostility.

4.1   Greek Astronomy

References to astronomical knowledge can be found also in the text of Homer's *Odyssey* when the poet cites some constellations like the Big Dipper or Orion or the Pleiades cluster, and he describes how the stars were used for navigation. Other examples are in the works of Hesiod, where he reports information that can be used to determine the best time for plowing, sowing and reaping. Thales of Miletus and Pythagoras of Samos are philosophers who provided significant scientific contributions, but no written documents about this subject have survived to our age, and the story that Thales correctly predicted the total eclipse of the Sun on 28 May 585 BC has probably to be regarded as a legend. In 450 BC the Greeks began to successfully study the motion of planets and in the fifth century BC Philolaus was a supporter of the Pythagorean theory, proposing that the Earth, the Sun, Moon and planets moved around a central focus, the Earth having a full orbit around the fire every 24 h which explained the daily motion of the Sun and the stars. About a century later, in 370 BC, Eudoxus of Cnidus explained the motions of the stars and those of the Sun, the Moon, and the planets assuming that the stars were positioned on the inner surface of an enormous sphere revolving around the Earth in 24 h, and using other transparent spheres smaller than this, revolving with different directions and speeds, for the other celestial bodies.

This was at the origin of the geocentric astronomical model which was accepted for almost 2000 years, but Greeks produced other and different astronomical models. About 100 years later Aristarchus of Samos proposed that daily movement of the stars was due to the Earth's spin around its own axis once a day, and that our planet, like the others, was orbiting around the Sun. This heliocentric theory was to be revived only many centuries in the future, with Copernicus and Galileo Galilei. During the third century BC, instead, Apollonius of Perga introduced the concepts of Epycicles and Deferents in the original model of Eudoxus to explain phenomena like the retrograde motion of the planets and the varying speed of the Moon, while Ptolemy, who lived in the second century AD, further extended this model by using the concept of Equant which allowed him to explain the velocity variation in planetary motion, thus producing more accurate predictions of their ephemerides. To test their theories, the Greeks also carried out astronomical observations which were organized into celestial tables, producing the first "modern" stellar catalogs, like those of Hipparchus of Nicaea or Ptolemy, who lived in the second century AD, showing the location of over 1000 stars.

In later centuries the tradition of Greek Astronomy was kept alive by other philosophers. Among them we can find Hypatia, who lived in Alexandria of Egypt, during the early centuries of the Christian era. A follower of Plato, today she is considered the first important scientist of the West, but we will deal with her soon.

4.2 Works

The exceptional skills of Aglaonike were attributed to magic arts, rather than a scientific expertise, and probably being a woman further encouraged such beliefs, making even more incredible and extraordinary her abilities in a society where women were usually suppressed and could emerge only as priestess or sorceress. Therefore, although her skills came from actual scientific knowledge, she was rather considered by the people to be a sort of witch who could exercise power over others through the fear aroused by the apparent control of natural phenomena.

4.3 Curious Facts

In the year 77 AD, Pliny wrote: "For a long time a way has been found to predict chronological events in advance, not only those relating to the day and at night, but also the time of eclipses of the Sun and the Moon. Yet it is rooted in a large part of the people the primitive belief that these phenomena are caused by witchcraft and herbs, and that this science, the only one that belongs to the women, is more credible than anything else." (Naturalis Historia, XXV, 10)

4.4   As They Said of Her

She is the first woman astronomer mentioned in the poem "Letter from Caroline Herschel" by Siv Cedering.

The period in which Aglaonike actually lived is uncertain. She is dated back to the fifth century BC, according to some sources, or at around 200 BC according to others. Plutarch in *Conjugalia Praecepta* 48, 145c and *De Defectu oraculorum* 13, 417a and Apollonius of Rhodes in *Scholia* AR. 4.59 report that she was a daughter of Hegetor, or Hegemon.

Some sources refer to her as a sorceress who could make the Moon disappear at will, following a common and magic interpretation of lunar eclipses. In fact, due to the superstitions of the time, all those who were familiar with astronomical phenomena, such as the periods of the full moon and Eclipse cycles, were regarded by the people as persons who "make the Moon disappear." For this Aglaonike was also called "The Witch of Thessaly".

One of the craters on Venus is named after Aglaonike.

Chapter 5
Hypatia of Alexandria (355 or 370 ca. to 415)

She became much better than her teacher especially in the art of observation of the stars. (Philostorgius, historian, contemporary of Hypatia)

Beautiful, highly educated and wise, Hypatia was born in Alexandria, Egypt, in an era when women were not considered persons, but her father Theon, Rector of the Library of Alexandria, took care of her education to make her "a perfect human being".

Portrait of Hypatia drawn by Jules Maurice Gaspard (1862–1919)

© Springer International Publishing Switzerland 2016
G. Bernardi, *The Unforgotten Sisters*, Springer Praxis Books in Popular Astronomy,
DOI 10.1007/978-3-319-26127-0_5

She deepened her studies in Athens and in Italy and became a philosopher, a mathematician and an astronomer, succeeding her father at the age of 31 years at the helm of the most famous ancient Academy. A pagan and a convinced supporter of the distinction between religion and knowledge, Hypatia was killed in an ambush hatched by a group of Christian fanatics, linked to Bishop Cyril, then Patriarch of Alexandria, in 415 AD.

5.1 Alexandria of Egypt

"Alexandria is the flower of all the towns and is famous for many splendid monuments, due to his great genius founder and architect Dinocrates. [...]" So the city appears in the eyes of the traveler and historian Ammianus Marcellinus, who visited it between 363 and 366 AD. He sings of the healthy climate, the magnificent harbor with a Lighthouse, the temples, the statues, the colonnades comparable to those of Rome, but above all the libraries.

The city of Alexandria was designed and built in a very short time by Alexander the Great in 331 BC and ruled by the Ptolemies. The big eastern city harbor overlooked the Royal district: the Bruchion. Inside was the Museum with the Library attached to it, and was one of the most celebrated centers of scientific, literary and philosophical research of the Hellenistic world which, even in the fourth century AD, was keeping alive such old traditions.

Indeed the historian, in addition to his description of the city, leaves us a precious testimony of the scientific activity of Alexandria: "Here everything that lies hidden is brought to light by the geometric beam. Among them music is not entirely extinct yet, nor is harmony silent, and although rare, someone revived again the study of the movements of the world and of the stars, while others are learned in the science of numbers; moreover, a few are experts of the doctrine which indicates the ways of fate". So in the fourth century AD, Alexandria was still active not only in the study of geometry and mathematics, music, harmony, and astronomy, but also in astrology and divination.

5.2 Who Was Hypatia

Hypatia's father, Theon, was a mathematician, heir to one of the largest and most durable schools of all time: the mathematical school founded by Euclid in the beginning of the third century BC. He became best known for his astronomical studies and his presence at the mathematical school is attested to by reports of two eclipses which he observed in Alexandria in 364: the June 16th solar eclipse and a lunar eclipse on November 26.

In order to understand the context in which Hypatia and Theon were working, it has to be remembered that the Hellenistic school (of which Alexandria is the best known center) produced the largest corpus of scientific knowledge of the ancient world. This included fundamental mathematical works like Euclid's Elements or the Conics of Apollonius of Perga as well as the treatises of Archimedes or the Arithmetic of Diophantus, the greatest arithmetic work in the Greek world, which is at the origin of a type of mathematical research still studied as "Diophantine analysis". The Hellenistic geometry-based mathematics was largely applied to physical sciences, and in particular to Astronomy. The first known heliocentric hypothesis was made by Aristarchus. Hipparchus was one of the most renown astronomers of antiquity, and his tradition was partly recovered about 300 years later by Ptolemy, an Alexandrian astronomer, mathematician and geographer of the second century AD, author of the monumental compilation mathematics known as Mathematical System or Almagest, the geocentric theory which lasted throughout the Middle Ages until the "Copernican Revolution".

Over the centuries, these works needed to be updated; the so-called "commentaries": books that could pass this knowledge on to the next generations in an appropriate and intelligible way, with updates and checks based on the most recent observations.

Theon read, taught and commented all the Mathematical System of Ptolemy and was a specialist in the art of observation of the stars. He was the author of original works like "On the Nile flood" and "On the rising of the Dog", a Greeks' way to refer to Sirius, which the ancient Egyptians associated to the goddess Isis, and which was of fundamental importance for the Nile's annual flooding. In particular, in the third book of the Commentary of Theon of Ptolemy's Mathematician System, observations of the heliacal raising of this star (a star has an heliacal rising when it rises a few moments before the Sun) and of the Sothic's year (a particular Egyptian period of 1461 years). With regard to the commentaries, little of what he wrote survived to our days. Of one of them, the "Commentary on small astrolabe", we know almost just the title, while the "Commentary on Ptolemy's easy tables" is his only fully preserved work.

Hypatia was born in Alexandria shortly after the middle of the fourth century AD, in one of the most renowned families of the city, which gave her the opportunity to access knowledge despite her sex. Nothing is known about her mother, but we know that she had a brother named Epifanio because his father dedicated to him his small commentary on "The easy Tables", probably to introduce him to the study of astronomy, and it is likely that the IV book of the "Great commentary" was also dedicated to him.

Philostorgius, historian of the Church and contemporary of Hypatia, writes that she "learned from her father the mathematical sciences, but she became much better of her teacher in the art of observation of the stars." Other sources describe Hypatia as "of nobler nature of her father, she was not content with knowledge that comes through the mathematical sciences, to which he had introduced her, but not without elevation of mind, she devoted herself to the other philosophical sciences".

Hypatia, in fact, deepened her studies in Athens and in Italy. She was admired for her beauty and her wisdom, but she never married. At the age of 31 years, she succeeded her father in the role of guide and teacher of the scientific community of the famous Museum of Alexandria.

5.3   The Museum

The Museum of Alexandria was founded by Ptolemy Soter in 280 BC, becoming immediately an important center of cultural activity. Originally, given its association with the worship of the Muses, the Museum resembled very closely the organization of the ancient Academy of Plato, and of Aristotle's Lyceum, by which apparently it was inspired. It was headed by a priest, appointed by the Ptolemies, head of the religious aspect of the community and a President with administrative and supervision tasks of the funds of the institution.

Lessons were held mostly as informal discussions and conversations; debates, conferences and symposia were frequent and typical, accompanied by jokes, epigrams and problems proposed to the public, which included also the Ptolemaic Kings and Queens. The buildings, sumptuously furnished by the Ptolemies, included a common refectory, an exedra for debates and conferences and a "Peripaton" (colonnade) shaded by trees. Although there are no sources from which to make a confirmation, it is likely that lessons were also taken in these places. The Ptolemies built also two large libraries: the first was annexed to the Museum; the other, also called "daughter of the Museum", located in the Serapeum. The Internal Library, which was the oldest and largest, kept about four hundred thousand "compound" volumes, that is, containing one or more works of various authors, and ninety thousand "simple" volumes. The External Library, smaller, housed about forty thousand volumes. It was part of the broader museum complex in Alexandria, which also included the anatomy Amphitheater and the Observatory.

Artistic Rendering of the Library of Alexandria

The libraries and the Museum become the largest cultural centers of antiquity, nurturing the scientific activity of many people like Eratosthenes, one of the scientists who flourished in the third century BC in this establishment, who estimated the Earth's circumference with an error of only 70–40,000 km. Several scientists succeeded him in this school and the last was a woman: Hypatia. This structure also played a fundamental role in transmission of scientific knowledge and of all culture in general, and although their destruction is often attributed to the Arab conquest of the seventh century AD, it is now ascertained that this was just the final blow to a severely ruined complex. The first one had been given already in 150 BC by King Ptolemy VIII during the period of civil war and riots coinciding with the uprising influence of Rome in these regions. Then in 47/48 BC the Internal Library was badly damaged during the conflict with Julius Caesar, and about one-third of the volumes were burned, while in 392 AD Christians burned many of the remaining ones.

5.4   Works

The researches and studies of Hypatia were strongly oriented towards teaching, transmission and commentary of ancient texts, in fact, the already cited historian Philostorgius says that Hypatia "introduced many to the mathematical sciences" and that "she became much better than her master in art of observation of the stars".

As regards the former, it is known that Hypatia taught in Alexandria continuously for more than 20 years: around 393 Synesius, a young man of the aristocracy of Cyrene (modern Libya), arrived in the Egyptian metropolis attracted by her fame; from that date the pedagogical activity of Hypatia is attested until the day of her assassination in March 415.

On the other side, the oldest testimony of Hypatia's scientific work is located in the same work of Theon who, in the header of the third book of his commentary on Ptolemy's Mathematical System, writes: "Comment by Theon of Alexandria on the third book of Ptolemy's Mathematical System. Controlled by the philosopher Hypatia, daughter of mine".

Hypatia wrote original works that, like those of her father, have disappeared: a commentary on Diophantus (the father of algebra) of 13 volumes, the Canon Astronomical (a collection of tables on celestial bodies), and a commentary on the Conics of Apollonius, a geometry treatise. It is deplorable that nothing remains but the titles of some of the works of a scientist so renowned in her time. However, these titles are clues that outline a theoretical trajectory, and our sources have handed down assurance that Hypatia wrote an original astronomical work. Apparently she had completed testing and observations that cannot simply be placed at the margin of Ptolemy's Mathematical System, already commented on by her father, but required instead a separate discussion.

Centuries of Church history, or perhaps it would be more correct to say "centuries of domination of ecclesiastical power domain", led many generations to the belief that the Ptolemaic System had said the final word on celestial geographies. It was during these centuries, in fact, that the Mathematical System started to be called (and known) under the title "Almagest" ("The greatest"), a term coined by the Arabs from the Greek in calling these compilations "The great system".

The main goal of the Alexandrian mathematical tradition was the study of motion of the stars, but this aim is rarely made explicit. Compared to this tradition, the title chosen by Hypatia (Astronomical Canon) represents an innovation, and it seems to indicate that the measures she carried out, and tested by herself and by those who worked with her, convinced her that it was possible to shift the point of view of the late Hellenistic science, putting more emphasis on the final application of her studies with respect to the traditional prominence given to the geometrical and mathematical aspects of such subjects. In this sense she seems to recover a practice already used by authors like Archimedes and Hipparchus.

What is known is that she commented the *Conics* of Apollonius and the *Arithmetic* of Diophantus, works that are, each in their field (respectively the geometry and the arithmetic), the highest expressions of the school of Alexandria.

In this way Hypatia combined two different and distinct interests: the classical geometry and this part of arithmetic developed by Diophantus, which was later called algebra.

These were separate and independent disciplines, and she was interested in the discovery of the laws of motion of the stars, so it appears natural to think that, with Diophantus, she tried to find new quantitative methods suitable for solving the difficulties that observation of the sky continued to give her. The application of arithmetic to geometry was an implicit exercise in the approach of science that Hypatia had learned through her father.

Since the third century BC, Apollonius, who today is considered the founder of quantitative mathematical astronomy, had shown the importance of arithmetic methods (which today we would say algebraic) applied to geometry. His work had effect on those of Hipparchus, Ptolemy, Theon and all astronomers who were included within the tradition outlined by these names. Compared to them, Hypatia seems to take advantage of the work in the field of number theory of Diophantus, who probably lived not earlier than the third century AD.

It is now possible to say what intuition had forced Hypatia to seek the help of the Arithmetic of Diophantus, nor what advantage she had then taken from this new and valuable tool. We do not know whether she moved in the same direction as Descartes and Fermat, more than a thousand years later, or instead if she saw a different direction that we cannot even imagine. All we can say with some certainty is what, once again, Philostorgius says: "*that Hypatia had discovered something new about the motion of the stars and that she made this new knowledge accessible to men and women of her time, explaining her new observations in an original work which she entitled Astronomical Canon*". It is not impossible that her work, especially if so innovative compared to that of Ptolemy, remained at the margin or even misunderstood in a school that, after Hypatia's death, retained only a marginal interest in astronomy.

Besides theoretical studies, Hypatia was apparently interested also in mechanics and applied technology, and in particular she is credited with the invention or the improving of two instruments: a hydrometer and a flat astrolabe.

The first tool was used to determine the specific weight of a liquid. It was designed as a sealed tube having a weight attached to one end which was left sinking in a liquid, whose specific gravity could be read on a graduated scale according to how much the tube sank. The improved astrolabe designed by Hypatia, instead, consisted of two perforated metal discs, rotating one above the other with a removable pin: it was used to calculate the time, to define the position of the Sun, of the stars and of the planets. It seems that using this tool she solved some problems of spherical astronomy. This information is supported in the writings of a valuable witness: her student Synesius who, in a letter (epistle 154 of "Synesius Epistles" addressed to Peon and Hypatia) described an astrolabe that he designed "based on what was taught by my revered teacher", starting from an intuition of Hipparchus, then neglected by Ptolemy and "by the divine ranks of his successors".

In the Epistle 15, always referring to the philosopher Hypatia, one can read the following meaningful request: "I am in such evil fortune that I need a hydroscope.

See that one is cast in brass for me and put together. The instrument in question is a cylindrical tube, which has the shape of an aulos (a sort of ancient flute) and is about the same size. It has notches in a perpendicular line, by means of which we are able to test the weight of the waters. A cone forms a lid at one of the extremities, closely fitted to the tube. The cone and the tube have one base only. This is called the baryllium. Whenever you place the tube in a liquid, it remains erect. You can then count the notches at your ease, and in this way ascertain the specific gravity of the water" (translation taken from http://www.livius.org/su-sz/synesius/synesius_let ter_015.html).

This reference shows, among other things, that at the school of Hypatia, Synesius had learned to "transfer the theory in the material world", as he says, and then to build tools that could help to prove on an experimental basis the theories of their ancient predecessors, which might be interpreted as an attempt to revive the method of early Hellenistic science. This is certainly an intriguing hypothesis but, as mentioned above, unfortunately the complete loss of her works makes it impossible to put this statement on a solid basis.

5.5   The End

Although Hypatia lived in an era when women were considered inferior beings, she was so celebrated for her knowledge and wisdom that many faced long journeys to attend her lectures. She did not share her knowledge with her students only, but historians report that "the woman, throwing herself on the mantle and venturing out in the middle of the city, explained publicly, to anyone who wanted to listen, Plato or Aristotle or the works of any other philosopher. [. . .] Since that was the nature of Hypatia, ready to the discussion in speeches, and a shrewd politician in the acting, the rest of the city rightly loved and respected her, and the leaders, every time she was available to take charge of public issues, used to first go to her."

She however lived in a dangerous epoch, and in a city where the strong rivalry and confrontation between different parties often lead to violent riots. One of these parties was that of the Christians, led by Cyril, Bishop of Alexandria, who began a persecution against the neo-Platonists "heretical" who basically were the pagan high society allied with the Roman governor Orestes, and the Jews. It is not known if Hypatia took an active part in this battle for power, but she was certainly a prominent exponent of the "philosophers" faction, and in this respect a potential target for the Christians.

"Death of the philosopher Hypatia, in Alexandria". This particular version is from the book Vies des savants illustres, depuis l'antiquité jusqu'au dix-neuvième siècle , by Louis Figuier, first published in 1866

Now all sources agree on what actually happened, but they differ on how they recall the facts. Indeed, according to the Christian philosopher Socrates Scholasticus, contemporary of Hypatia, the philosopher was attacked and brutally killed in the streets of Alexandria by a mob of Christian fanatics who then raged on and finally burned her corpse. While confirming the mere facts, the pagan philosopher Damascius instead, living about one century after, explains the assassination of Hypatia as a conspiracy organized by Cyril, who envied her fame and was afraid of her influence over the city. According to this later version, it was Cyril himself who sent the mob, ordering the murder of Hypatia in 415 AD.

Whatever the truth is, her assassination marked the end and the dispersion of the school that made reference to the Greek philosopher, who in fact is remembered in history as the last great Alexandrian astronomer. Apparently Cyril wanted to destroy everything, even cancelling the memory of the mathematical astronomer of Alexandria. The little that was saved was plundered by the Crusaders in the library of Constantinople (Istanbul) and today it is preserved in Rome, in the Vatican Library.

5.6   As They Said of Her

Philostorgius, historian of the Church and contemporary of Hypatia, writes that she "learned from her father the mathematical sciences, but she became much better of her teacher especially in the art of observation of the stars" and "introduced many to the mathematical sciences."

Other sources describe her as "of more noble nature than the father, was not content of the knowledge that comes through mathematical sciences to which she had been introduced by him but, not without elevation of mind, she devoted herself to the other philosophical sciences."

The earliest testimony of the scientific activity of Hypatia is in the same work of Theon; in the header of the third book of his commentary on Ptolemy's Mathematical System, he writes: "Theon of Alexandria's commentary on the third book of Ptolemy's Mathematical System. Edition controlled by the philosopher Hypatia, daughter of mine."

As already remembered, Philostorgius also let us know: "that Hypatia had discovered something new about the motion of the stars and that she made this new knowledge accessible to men and women of her time, explaining her new observations in an original work entitled Astronomical Canon."

The contemporary Christian historiographer Socrates Scholasticus, in *Ecclesiastical History*, gives the following description of Hypatia: "There was a woman in Alexandria named Hypatia, daughter of the philosopher Theon, who made such attainments in literature and science, as to far surpass all the philosophers of her own time. Having succeeded to the school of Plato and Plotinus, she explained the principles of philosophy to her auditors, many of whom came from a distance to enjoy her learning. On account of the self-possession and ease of manner which she had acquired in consequence of the cultivation of her mind, she not infrequently appeared in public in the presence of the magistrates. Neither did she feel abashed in going to an assembly of men, for everybody, on account of her extraordinary dignity and virtue, admired her the more."

5.7   Curious Facts

On 16 October, 2002 the new library of Alexandria of Egypt, or *Bibliotheca Alexandrina*, was officially inaugurated. Built close to the place where the legendary original one was erected, similarly to its predecessor, it is part of the academic structure of the city, and the complex includes also other structures like art galleries and a planetarium.

On June 27th, St. Cyril of Alexandria, doctor of the Church, is celebrated.

Chapter 6
Sonduk (?–647)

Will I ever know the truth about the stars?
I'm too young to engage in theories about our Universe.
I just know that I want to understand more. I want to know all
I can. Why should it be forbidden?

Let us now leave the Mediterranean area, moving to the Far East, in Korea. Here was born in 610 AD Princess Sonduk or Sondok or Seondeok of the Silla Dynasty. The above introductory sentence had been filed by the same Princess, when she was 15 years old, in a votive jar dedicated to her grandmother. She later became the first female monarch of Korea, ruling her country for 14 years. Her father, Jinpyeong, was the king of the Silla kingdom, that was born as a city-state in 57 BC and, after having had successively emerged as a kingdom in about 350 AD, by the end of the seventh century had managed to unify the whole peninsula. Having no sons, he chose as his heir his daughter Sonduk, which was no great surprise for a number of reasons: one was that women in this period had a certain degree of influence already as advisers, queen dowagers, and regents. Women were also heads of families, since matrilineal lines of descent existed alongside patrilineal lines, and the Confucian model, which placed women in a subordinate position within the family, was not to have a major impact in Korea until the fifteenth century, so that during the Silla kingdom, women's status remained relatively high. Sonduk was chosen by the king also because of her brilliant mind, a specific trait that had become evident early in her life. One anecdote is about a box of peony seeds from China accompanied by a painting of what the flowers looked like that the king had received. Looking at the picture, 7 year old Sonduk remarked that while the flower was pretty it was a pity that it did not smell: "If it did, there would be butterflies and bees around the flower in the painting." Her observation about the peonies lack of smell proved correct, one illustration among many of her intelligence, and thus ability to rule.

As a very young lady she found herself attracted to the study of the stars since her tutor, the Chinese ambassador, was also an astronomer who, however, did not believe that astronomy was suitable for a woman. This link with the Chinese ambassador probably was a kind of political one, since Sonduk was to succeed

© Springer International Publishing Switzerland 2016
G. Bernardi, *The Unforgotten Sisters*, Springer Praxis Books in Popular Astronomy,
DOI 10.1007/978-3-319-26127-0_6

her father as Queen of Silla, who was a strategic ally of the Tang dynasty in the Korean peninsula, then divided into fighting states. She eventually became the sole ruler of Silla in 632 AD or, according to other scholars, in 634 AD and she ruled until 647 AD, dying on 17 February. She was the first of three females rulers of the kingdom, and was then succeeded by her cousin Chindok, who ruled until 654 AD.

Sonduk's reign was a violent one; rebellions and fighting in the neighboring kingdom of Paekche (which eventually became part of the unified Korea under the Silla kingdom thanks to the help of the Chinese Tang dynasty) filled her days. Yet, in her 16 (or 14) years as queen of Korea, her wit was to her advantage; she kept the kingdom together and extended its ties to China, sending scholars to learn from that august kingdom. Like China's Empress Wu Zetian, Sonduk was drawn to Buddhism and presided over the completion of Buddhist temples.

Her main contribution to astronomy, among those remembered by history, was the construction of the Astronomical Observatory called the Tower of the Moon and the Stars, considered the first Observatory in the Far East. Sonduk begged her father for several years to start the work, but she eventually succeeded, and its ruins can still be visited even today. Located in Kyongju about 100 km north of Pusan, in South Korea, home to the ancient capital of his dynasty, the tower was built with stone blocks placed on 27 levels, possibly as a reference to the Queen who was the twenty-seventh ruler of Silla. It reaches about nine feet tall and still today challenges the centuries, winning the primacy of the oldest astronomical observatory in Asia.

The "Star-Gazing Tower," or Cheomseongdae, considered the first dedicated observatory in the Far East. The tower built by Sonduk, still stands in the old Silla capital of Gyeongju, South Korea. Creative Commons Attribution 2.0 Generic

6.1 Korean Astronomy

Korean astronomers studied the positions and the apparent motion of celestial bodies. They filled out the annual calendar and predicted celestial phenomena, such as the spring Equinox and autumn, the winter Solstice and summer, and solar and lunar eclipses. They did not forget to observe comets and meteors, taking care to record the apparitions, because all this for them was extremely important. The King, in fact, ruled over the Korean people with the power that they thought he had sent from above and as representative, he served the heavens and kept close contact with them in order to lead his people on the right track. It was thought that taking note of changes in the heavens, they were able to predict the future, so there was a special astronomical Bureau under the orders of the Royal Palace, with this very important assignment. The astronomy in this manner appears more a religion than a science. The ancient astronomers employed also armillary spheres and astrolabes to know and better understand the stars, previous tools, for example, were used to calculate the time when stood or was setting the astro.

6.2   Works

As already said, Queen Sonduk's main contribution to astronomy was construction of a tower for astronomical observations, called Chonsongdae (see figure); it is one of the oldest existing structures in Korea and the oldest existing complex of this kind in Asia. Doubts were cast on the interpretation of this building as an astronomical observatory, but a study done in 2001 apparently settled the debate. It was in fact discovered that the number of observations reported in the historical records during the three centuries after the construction of the tower until the fall of the Silla kingdom was more than doubled with respect to those reported for the seven centuries preceding its erection. Still there is already an uncertainty about the duration of the works. Historical documents report that it was built between 633 and 647 AD, yet this period seems quite long for this kind of structure. The tower consists of 365 stones, corresponding to the days of the year, arranged in 27 circular sections each 30 cm high. As already said, it is supposed that the twenty-seven sections can symbolize the 27th sovereign, Sonduk, but they could also be attributed to the major known constellations. The base is formed by 12 rectangular stones, like the months of the year, and its square shape along with the cylindrical one of the standing tower with tapered tip is believed to have a symbolic meaning. Actually they could represent the ancient popular belief about a round heaven surrounding a squared Earth. Korean astronomers carried out studies on the positions and apparent motions of celestial bodies, compiled the annual calendar and made forecasts on celestial phenomena such as the Spring and Autumn Equinox, the Winter and Summer Solstice, as well as predicting solar and lunar eclipses. In addition they also recorded the passage of comets or swarms of meteorites. The square opening that can be seen at mid-height is facing South. It was created to allow sunlight to fall onto a point on the floor at the Spring and Autumn Equinox, and when the Sun crossed the Meridian. The window was designed in such a way that it did not receive the light of the Sun at the Summer Solstice, so the tower was used as a calendar, accurately indicating the passage of the four seasons.

6.3   Curious Facts

The name Chomsongdae literally means "tower for stargazing" and, although built 1400 years ago, this stone structure remains mostly intact. It is considered the oldest Astronomical Observatory of its kind in the world, but its role was recognized only a century ago. Apparently, its scientific value started to be appreciated only after Korea entered into the modern world. The only entrance to the tower, probably accessible from an external ladder or staircase, is a square window placed at about four meters from the ground. According to the historical accounts, when astronomers had to observe in the Sonduk's tower they laid on their back and watched the celestial objects through four domes arranged in a square and oriented towards the four cardinal points.

In ancient Korea women enjoyed a respectful condition as female shamans, and it may well have been the case that Sonduk's respect as a ruler was reinforced by such tradition. Actually the word shaman was assumed to apply to women, to whom great powers were recognized as intermediaries between gods and humans. Some presided over national ceremonies, but they were mostly a kind of family priestess, whose role was usually inherited. Their powers were exerted through spirit possession, which allowed the shamans to perform, among other powers, healings and exorcisms, to reveal causes of family strife and advise on their resolution. As foretellers, enormous power was attributed to shamans, and it is reported that Sonduk was revered for her ability to anticipate advents. No further detail about the kind of forecasts she was able to provide is known, but it is not difficult to suppose that this ability may be linked to her interest in astronomy, since many other cases in history are known where astronomical forecasts had been exploited to reinforce the prestige and the influence of the foreteller.

6.4 As They Said of Her

The young Sonduk became interested in the study of the stars and made observations every night. She was mostly self-taught, although she received some education by the Royal Astronomers. At the age of 15 she studied Confucianism with the Chinese Ambassador Lin Fang, who was also an astronomer, and had submitted a new official calendar to the King, Sonduk's father, convincing him that the Chinese calendar was better than the Korean. Sonduk wished to discuss astronomy with the Chinese ambassador, who however believed that female occupation had to be kept within the home. Indeed he replied: "Surely you can't think I can have a conversation on such important topics with a young woman! It would be unnatural and totally inappropriate". Despite this "encouragement", during a solar eclipse that occurred in Korea, the young Princess showed that her knowledge was more than enough to discuss such matters with him on even grounds, as she was able to predict the event and its duration with high accuracy. This angered the Ambassador who said: "Astronomy is not for women, do anything feminine such as care of silkworms!", and eventually managed to convince her father to preclude Sonduk from any further study of the stars.

Part II
Timelines from Fatima to Jeanne Dumée

800 birth of the Holy Roman Empire

1096 the first Crusade

1270 travels of Marco Polo

1286 first witnesses of eyeglasses

1440 invention of printing

1492 discovery of the Americas

1543 publication of Copernicus's "*De revolutionibus orbium coelestium*" and Heliocentric Theory

1609 Galileo's telescope

1619 Kepler's third law

1648 end of the Thirty Years' War

1655 Huygens discovers the rings of Saturn

1658 Huygens perfects the pendulum clock

1660 foundation of the Royal Society as the first scientific society

1676 Roemer measures the speed of light

1682 Halley discovers the Comet that bears his name

1684 Leibniz publishes the first mathematical work on differential calculus that will have important repercussions for astronomy

1687 Newton publishes his findings on the laws of dynamics and of gravity

Chapter 7
Fátima of Madrid (Tenth Century)

The Mistery of Fátima

A very interesting "case" is that of Fátima of Madrid. Such an Arab name associated to the capital of modern Spain should not be surprising once we know that she was an Andalusian astronomer who lived in the tenth century in Spain which at that time was called Al-Andalus. Indeed, in this historical period this territory had been ruled for more than two centuries by Muslims, as the outcome of the Muslim conquest of the Visigoths' Hispania. The tenth century witnessed the magnificence of the Caliphate of Córdoba (929–1031) under which Spain was a beacon of learning, with its capital playing the role of a leading cultural and economic center in Europe as well as in the Islamic world. Like other Muslim countries of those centuries, the importance of Al-Andalus in the history of science came not only from its achievements in trigonometry, astronomy and other scientific fields, but also from its role as a conduit for culture and science between the Islamic and Christian worlds, which allowed the renaissance of science in Europe after its eclipse during the early Medieval Age. It is in such favorable environment that this woman could work, in close collaboration with her father, the Islamic astronomer and scientist al-Maslamah Mayriti, whose name means "man from Madrid" and thus known as "El Madrileno".

7.1 Arab and Islamic Astronomy

While Western civilization was experiencing a period of stasis, if not decline, of its early Medieval Age, the Islamic Empire, after its fast expansion of the seventh and eighth centuries, stretched from central Asia to southern Europe, holding a leading role in military power as well as in cultural and scientific knowledge.

© Springer International Publishing Switzerland 2016   45
G. Bernardi, *The Unforgotten Sisters*, Springer Praxis Books in Popular Astronomy,
DOI 10.1007/978-3-319-26127-0_7

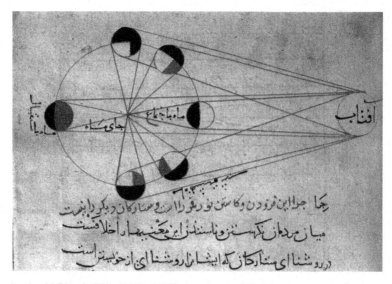

Illustration by Al-Biruni (973–1048) of different phases of the Moon

The latter was favored by the translation into Arabic of many classical Greek and Roman. For instance, Ptolemy's model of an earth-centered universe formed the basis of Arab and Islamic astronomy, which in this way was later transmitted to Europe, but Islamic astronomers were not only mere translators. From a theoretical point of view, Ptolemy's geocentric model was critically investigated and questioned, especially in regard of his Equant technique. At the same time they were also active as accurate observers, perhaps the most famous of whom is Ulugh Beg, who lived in the fifteenth century. This spurred parallel advances on the technological side, by the development of refined astronomical instruments. Possibly the most famous example is the astrolabe, a sort of analogical calculator used to calculate and predict the positions of stars and planets, but also the local time and latitude. Invented by Greeks, it was adopted and perfected by the Arabs, remaining the most accurate astronomical instrument for many centuries.

It is then not surprising to discover that a large extent of Arab astronomical legacy survived to the present days. This heritage, for instance, can be found in a significant number of stars in the sky, such as Aldebaran and Altair, but also many astronomical terms such as alidade, azimuth, and almucantar, are still referred to by their Arabic names. The interest of the Islamic world for astronomy is also testified by a large corpus of literature still, numbering approximately 10,000 manuscripts, which is still lasting scattered throughout the world.

One peculiar characteristic of the Muslims' interest for astronomy rests in its use for fulfilling practical religious needs. These ranged from establishing the correct moments of the daily praying to the starting of the religious festivals, but they also included the ability of determining the direction of Mecca, Muslim sacred city, from any geographical position.

7.2   Works

Fátima worked with her father on astronomical and mathematical investigations and wrote several treatises on astronomy, known as *"Corrections form Fátima"* that is her most famous work. Again in collaboration with her father in editing the *"Astronomical tables of al-Khwarizmi"*, correcting it for the Meridian passing through Córdoba. Together they worked on zijes, that is Arabic tables including calendars, ephemerides of the planets, the Sun, the Moon, with their eclipses and information on the visibility of the Moon, but also trigonometric and spherical astronomy tables and other information. She knew, in addition to Spanish and Arabic, also other languages such as Hebrew, Greek and Latin. Finally, she worked with her father on a book entitled *"Astrolabe Treaty"*, on the use of this instrument, which is preserved in the Library of the Monastery of the Escorial, near Madrid.

7.3   Curious Facts and as They Said of Her

Despite the credit allowed to this character by the inclusion of her name among the distinguished female astronomers of the past during the International Year of Astronomy in 2009, or maybe exactly for this reason, recent claims strongly questioned the existence of Fátima of Madrid.

Information on her life appear in the edition of 1924 of the "Encyclopedia Universal Ilustrada Europeo Americana" of Espasa-Calpe which so far is the most ancient source on this character, since the references used to write this voice was not cited and are not known. This is one of the reasons why some specialists believe that she is not more than an invention. The historian of mathematics Angelo Requena Fraile says: *"Bridging the gap in time and space, our Espasa has fulfilled admirably with his role as solid and reliable resource. But in all such ambitious projects there are typos, crooks, or careless copyists, and unfortunately in science history sneaked one of those rogues. This is how a few myths, not documented nor confirmed by other sources, referring to characters that should not be found in the history but as myths, get created. The father of the history of science in Spain - Francisco Vera - already put us on guard when inquired about a geometrician, Bishop of Calahorra of the Visigothic era, called Luciniano that did not appear anywhere else than in the Espasa. Whit Fatima, the learned daughter of Maslama Madrid, we find something similar to what happens with Luciniano. There is no other reference, there are no other reliable sources we can count on."*

The Arabist Manuela Marin, specialist in history and biographies of el-Andalus holds the same thesis and, speaking of Maslama al-Mayriti, she says: *"He is, therefore, the most illustrious of Madrilenian of Andalus, and there is no need to invent, as it was done, a historically nonexistent daughter of him, which has been called "Fatima de Madrid" and that has surprisingly been included in the calendar 'Astronomers who made history', published on the occasion of the International*

Year of Astronomy (2009). The error, which comes from an old edition of the Espasa encyclopedia, has been reproducing uncritically, it is high time to stop it, although abundant circulation over the Internet does not give much hope in this regard."

Chapter 8
Hildegard von Bingen (1098–1179)

One feather in the wind abandoned of the trust of God (The Sibyl of the Rhine as self-designed)

After the tragic death of Hypatia of Alexandria, the school also faded and dispersed, after which the pursuit of knowledge passed to ecclesial institutions. Afterwards we have no notices of female scholars or educated women until the ninth century, which in any case came from aristocratic or royal families placed in monasteries. In such a sphere we can find eminent Abbesses, the most notable of which is Hildegard of Bingen, a German Benedictine nun who lived in the twelfth century and the most famous among the religious and medieval female scientists. She was born in the summer of 1098 in Bermersheim, near Alzey, in the now German region of Saxony, at the time part of the Holy Roman Empire. She was the last of ten children of Mechtild of Merxheim-Nahet and Hildebert of Bermersheim, an aristocratic family, and was sickly for much of her childhood. At the age of eight Hildegard was locked up in a convent at Disibodenberg led by an older nun named Jutta (in some sources cited as her aunt) who was the daughter of Count Stephan II of Sponheim. Here she had the opportunity to study and work and later to write several treatises on different subjects ranging from medicine to cosmology. Hildegard rose through the ranks of the Church. In 1136, upon Jutta's death, Hildegard was unanimously elected by her fellow nuns as "magistra", which is the Latin word for a female teacher or mistress. She went so far as to convince the Church to take the unusual step of allowing her to found two monasteries: Rupertsberg in 1150 and Eibingen in 1165. The first known female composer of sacred music and a painter, she also wrote works on cosmology that are contained in two books which are very good representatives of her specific way of writing. Actually, Hildegard wrote these works as visions coming from divine inspiration, describing their origin in this way: *"I spoke and wrote these things not by the invention of my heart or that of any other person, but as by the secret mysteries of God"*.

© Springer International Publishing Switzerland 2016
G. Bernardi, *The Unforgotten Sisters*, Springer Praxis Books in Popular Astronomy,
DOI 10.1007/978-3-319-26127-0_8

Hildegard of Bingen receiving a divine inspiration

In an epoch where science and theology were intertwined at a fundamental level, and females' knowledge was hardly accepted, or in any case confined to very specific fields and practices, this was of great help for spreading her theses and giving greater credibility to her writings. Sickly from birth, in her *Vita* Hildegard states that since a very young age she had experienced visions, and all of her works were illustrated by very detailed explanatory miniatures in which she portrayed herself in a corner in the act of observing "her visions".

8.1   Works

Among the works of Hildegard three of them, written in her characteristic "visions-inspired" style, are the most significant ones:

- *Liber Scivias* (Know the Ways) composed between 1142 and 1151,
- *Liber Vitae Meritorum* (Book of Life's Merits or Book of the Rewards of Life) composed between 1158 and 1163;
- *Liber Divinorum Operum* (Book of Divine Works, also known as *De operatione Dei*, On God's Activity, composed between 1163/1164–1172 and 1174).

In two out of these three, the *Liber Scivias* and the *Liber divinorum operum simplicis hominis*, are contained her cosmological models and their theological interpretation. In this way, in the twelfth century many of the cosmological ideas of the Judeo-Christian and Greek traditions, were circulated by Hildegard who, however, added many original concepts to the traditional forms of such models.

The Universal Man, Liber Divinorum Operum of St. Hildegard of Bingen, 1165. Copy of the thirteenth century

Her concept of cosmic structure can be summarized in the vision of a spherical Earth surrounded by concentric shells carrying bodies which can affect human events, an idea that dates back at least to the Pythagoreans, presented by Hildegard as a new revelation with unique details. In the *Liber Scivias* the Earth is made up of the four elements and is surrounded by a spherical atmosphere, called "Alba pellis" or "Aer lucidus." Each of the four primordial casings of the universe contains one of the cardinal winds, represented as the breath of a supernatural being and other, secondary winds. The first of these enclosures is a spherical water area, or "Aer aquosus," whose ends are placed outside the clouds, which contracts, expands and spreads, concealing or revealing the heavenly bodies below. In drawing this vision, the East is placed by Hildegard on top and the north on the left with an elongated east–west axis, so that the outside areas are egg-shaped. This is in contrast to the common model of many contemporary scholars who, reporting the original ideas of the ancients, conceived of a totally spherical universe. The "Purus aether" surrounds with its oval structure the Aer aquosus and is the largest of the primordial casings: it contains the Moon and the inner planets (Mercury and Venus) and the constellations of the fixed stars. We can find then the dark, thin belt of an inner fire called "Umbrosa pellis" or "Ignis niger" which is a source of light and heat. Finally, all is contained in the last casing, called "Lucidus ignis" or external fire, whose East end (the upper one) is elongated and pointed, and where the Sun and the outer planets (Mars, Jupiter and Saturn) are located. According to scientists, the weather and the seasons of the two terrestrial hemispheres are antithetical and the motion of the celestial spheres and the change of seasons on Earth are facilitated by the winds each casing. In practice, the prevailing winds act as a driving force for the lengthen of the day in the spring and its shorting in autumn:

> I looked and admired eastern winds and south, with their reactions, moved the firmament with the force of the breath and caused the movement from East to West over the Earth; and also the winds of the West and the North, with their side, received the impulse and addressed their gusts to dismiss it back from West to East [. . .] Also I saw that, while the days began to grow, the southern wind and its effects gradually lifted the firmament in the southern hemisphere to the North until the days do not cease to grow. Then, when the days began to get shorter, the Northern wind with its side portrayed by the brilliance of the Sun and gradually brought the firmament southward until, due to the lengthening of the day, the South wind began to bring it back up again.

In the second and third vision of *Liber dominorum*, Hildegard attributes to each wind qualities associated to various animals. Realizing that the planets were to have an additional movement independent from that of their spheres and different from that of the stars, and which had to include west to east motions, she added a new vision that reveals the existence of a force represented as a supernatural creature with a human face located in the external sphere of fire. Such force moved the planets from west to east, in the opposite direction to the motion of the firmament. Every sphere, every astronomical body, the winds and the clouds emit influences, represented by lines, towards a human figure: the microcosm. The doctrine of the macrocosm and the microcosm, as a cosmological theory that lasted throughout the

Renaissance, signed by scientists like Paracelsus, Harvey, Robert Boyle and Leibniz, was the central dogma of medieval science. It was based on the similarity between the structure of the universe and the human anatomy, where Hildegard included as well the qualities of the soul. The Abbess was trying to incorporate in her diagram of the macrocosm anatomical and physiological knowledge of the time, her idea of the human mind and her theological beliefs. Celestial elements influenced the human body by acting through the atmosphere on the blood and humors. Each cardinal wind was representative of the primordial zone where it originated and exerted its influence on the corresponding humor of the human body. With theories of such kind, it is easy to indulge in astrology, a discipline already discussed in the twelfth century, as today. While condemning it, Hildegard argued that the heavenly bodies could occasionally reveal divine signs.

In addition to these cosmological works, she wrote theological, botanical, medicinal texts, letters, poems and was also an accomplished composer and about 80 of her musical compositions have survived to our days; one of these, the *Ordo Virtutum* (Play of the Virtues) is an early example of liturgical drama.

Her writings and compositions reveal Hildegard's use of one form of modified medieval Latin which encompassed many invented, conflated and abridged words. She also used an alternative alphabet called *Lingua Ignota* (unknown language), which some scholars believe was also used to increase solidarity among her nuns. For these inventions she has been regarded as a sort of medieval precursor of the so-called "artificial" or "constructed" languages, and indeed she is patron saint of the Esperantists.

Hildegard of Bingen died in Bingen am Rhein on September 17, 1179 at the age of 81 years and Pope Benedict XVI, on 7 October 2012, named her a Doctor of the Church.

8.2 Curious Facts

Hildegard, with respect to the period in which she lived, was an unconventional and nonconformist nun. Ill-healthed but with a very strong character, at the age of three, as she reported, she first saw "The Shade of the Living Light". At the age of five she began to understand that she was experiencing visions but, although considering them as a gift, she was hesitant to speak of them to others, sharing later her experiences only to nun Jutta, who was also a visionary. In 1141 she received a vision that she interpreted as an instruction from God, to "write down that which you see and hear" and in her first theological text, *Scivas*, she describes her struggle:

> But I, though I saw and heard these things, refused to write for a long time through doubt and bad opinion and the diversity of human words, not with stubbornness but in the exercise of humility, until, laid low by the scourge of God, I fell upon a bed of sickness; then, compelled at last by many illnesses, and by the witness of a certain noble maiden of good conduct [the nun Richardis von Stade] and of that man whom I had secretly sought and found, as mentioned above, I set my hand to the writing. While I was doing it, I sensed, as I

mentioned before, the deep profundity of scriptural exposition; and, raising myself from
illness by the strength I received, I brought this work to a close – though just barely – in ten
years. (. . .) And I spoke and wrote these things not by the invention of my heart or that of
any other person, but as by the secret mysteries of God I heard and received them in the
heavenly places. And again I heard a voice from Heaven saying to me, 'Cry out therefore,
and write thus!'

Hildegard was afraid to leave the convent to confer with Bishops and Abbots,
Nobles and Princes, but at the same time, while in epistolary contact with the
Cistercian monk Bernard of Clairvaux, was not afraid to challenge with harsh words
the Emperor Frederick Barbarossa, until then her protector, when these two
opposed the legitimate Pope Alexander II. The emperor did not retaliate for the
affront, but dropped the friendship that until then had bound them.

In 1169 she succeeded in an exorcism on a woman called Sigewize, who had
been hospitalized in her convent, after that other monks previously tried and failed.
In a role quite unusual for a woman, she led personally the rite, although in presence
of seven male priests.

Although Hildegard was very much a product of her age and entertained a dim
view of sex, she was well ahead of her time in her appreciation and recognition of
the importance of sexual gratification for women. Despite presumably being a
virgin herself she may well be the first European to describe the female orgasm.

When Hildegard was dying, her sisters claimed they saw two streams of light
appear in the skies and cross over the room.

8.3   As They Said of Her

Dubbed "The Sibyl of the Rhine" for her prophecies, her strong-willed tempera-
ment was encouraged by her secure political position, which allowed her to become
engaged in the political and cultural field. Her essays were read at the Council of
Trent of 1147 by Pope Eugenius III, which called them true prophecies and
encouraged her to continue writing. As an aristocratic abbess, Hildegard repeatedly
described herself as "A *Feather* on the Breath of *God*." Faithful to the meaning of
her name, that is "protector of battles," she made her religion a weapon for a battle
that she led throughout her whole life, shaking the hearts and minds of her times.

Chapter 9
Sophie Brahe (1556–1643)

*Tycho Brahe's sister had exceptional knowledge in mathematics
and astronomy (Pierre Gassendi in the biography of Tycho
Brahe)*

She was considered one of the most erudite women of her time. Sophie Brahe was
the youngest of ten children of Otto Brahe (1518–1571) a State Councilor and Beate
Bille (1526–1605). Self-taught and sister of the famous astronomer Tycho, she
became his assistant on the island of Hven, which hosted his personal Astronomical
Observatory, supported by King Frederick II of Denmark. Indeed, she might have
been much more than a simple collaborator, as it is possible that she contributed to
the cosmological model known as *Tychonic*, after the name of the Danish astronomer. She also had a rich private life: she was married twice, and her varied
interests ranged from horticulture to medicine, and from alchemy to history.

© Springer International Publishing Switzerland 2016
G. Bernardi, *The Unforgotten Sisters*, Springer Praxis Books in Popular Astronomy,
DOI 10.1007/978-3-319-26127-0_9

Sophie Brahe, sister of Tycho Brahe, from P. Hansens "Illustreret Dansk Litteraturhistorie"

As a member of a wealthy family, she received a private education, as was customary at the time and suited to her rank. She studied German and Latin, but much of the intellectual interests that developed were linked to those of her elder brother Tycho, who, after studying law and philosophy at the Universities of Copenhagen and Leipzig, was supposed to follow a diplomatic career. He instead devoted himself entirely to the study of the stars, one of the disciplines that joined them. Destiny began to favor Sophie during her brother's return from abroad because of the worsening health conditions of their father. This event allowed her, at the age of ten, to study alone in the books that her brother brought home. Eventually, she turned out to be almost a child prodigy and this did not go unnoticed to Tycho who, because of her natural talent for scientific disciplines and of her growing curiosity, decided to care personally for the education of Sophie in subjects such as chemistry, Latin (the contemporary international language, especially among learned men), astronomy, alchemy and classical literature.

This is quite a meaningful fact for a person famous for his bad and irascible temper, sometimes even ruthless, that made him disliked among his servants and his assistants as well. Tycho found in Sophie a person with whom he could share his love for natural sciences. Such a situation calls to mind another famous pair of brother astronomers of the eighteenth century: William and Caroline Herschel, although the reasons that motivated the latter to take astronomical studies were very different.

9.1 Works

At the age of 14 Sophie was already an assistant of Tycho at Kunstorp. We can find her next to her brother during the observation of the Lunar Eclipse on 8 December 1573, which previously they had calculated together. The previous year, however, was crucial, since on the evening of November 11th 1572, a new star had appeared in the sky, in Cassiopeia. The strange celestial phenomenon remained visible for 18 months, during which it was studied and later described in the first work authored by Tycho and titled *De nova stella* or the new star. The astronomical event that had so attracted the attention of Brahe, was nothing other than a "nova" or a star that violently increases its brightness. This spectacular and mysterious event could not be understood in the framework of the Ptolemaic planetary model, made of fixed and immutable stars, and therefore added to the reasons that invited discussion.

The brothers were not followers of the Copernican model, alternative to the previous one, but proposed instead a model of the universe which was in part geocentric and in part heliocentric, where the Sun and the Moon orbit around the Earth while the other five planets (Mercury, Venus, Mars, Jupiter, Saturn) revolve around the Sun. This system became known as Tychonic, but unfortunately it is not possible to evaluate the extent of Sophie's contribution because, as it was natural at that time, everything is implicitly attributed to her brother. Nonetheless, there are many hints which suggest that she played an important role in the definition of the Tychonic model. First of all, Tycho was used to define Sophie as Urania, or his inspiring Muse. More significantly, he also called himself more an engineer than an astronomer, because he was extremely involved in the practical aspects of the making and improvement of their astronomical instrumentation, which allowed him to perform the most precise measurements of the time. He was also a pioneer in the field of observational astronomy, being one of the first men to stress the importance of a systematic and well-organized observation program. On the other hand, it seems that his sister was the more versed of the two on the mathematical and theoretical side of their work. As said, the Tychonic model was a hybrid between the geocentric and heliocentric model, and it did not have much luck in the circle of specialists, but it was one of the first to clearly break with the tradition of the material character of celestial spheres that had established itself with the medieval interpretation of the Ptolemaic model. This is in fact clearly incompatible with the intersecting orbits of several celestial bodies of this model.

Sophie Brahe married Otto Thott of Eriksholm and gave birth to one son, Tage, in 1580, but she returned to her family home after the death of her husband in 1588. There she plunged into her studies aside from the astronomical work, another effort where certainly Sophie took part was the compilation of the family tree of her family. It consists of over 900 handwritten pages about the noble families in Scandinavia. Published in 1626, it remains an important source for Danish historians today; the volume is kept at the Swedish University Library of Lund.

9.2 Projects

It seems that King of Denmark, Frederick II, had a debt of gratitude with an uncle of Tycho Brahe, who apparently saved him from a sure drowning. It was probably in part for this reason, but also because of the fame that Tycho had acquired in the field of astronomy, that he invited him to teach a class on astronomy at the University of Copenhagen. He also awarded him an annual pension, giving him even an entire island about 5 km long called Hven, near Copenhagen, on which was built the famous castle-observatory of Uraniborg, or the Town of Urania, and the Stjarneborg observatory, meaning "City of the stars". The sovereign's wife, Sophia, Queen of Denmark and Norway, was a patron of the Brahe brothers and at the death of her husband she retired from public life, devoting herself to astronomy and natural sciences.

The importance of the astronomical work conducted on this island by the Brahes is due to the fact that they were the first European astronomers to undertake regular and long-range observation program of the positions of fixed stars and of the planets. For this purpose, they used some of the most advanced instruments of the time, which had been designed and improved by themselves. The telescope had not been invented yet, but with their observations they were able to produce a catalogue of more than 1000 fixed stars and with the best accuracy of their epoch. They were also able to take into account the effect of atmospheric refraction on their observations, and some British researchers, who recently compared the star maps of Brahes with the current ones, discovered that their calculations were even more accurate than previously believed.

Their research was one of the first modern examples of an experimental work conducted in a rigorous and systematic way, which has to be compared with the widespread habit of their contemporary colleagues, who still employed the experimental data of the Ptolemaic era. But these observations, and therefore the work of Tycho and Sophie Brahe, are of paramount importance in the history of science, also because they nurtured one of the most fundamental scientific advancement of all times. After the death of Tycho in fact, Kepler had full access to the data collecting these precise observations, which until then the Danish astronomer had kept for his private use only, and thanks to them the German scientist was able to formulate his three laws on the planetary motion which, about 50 years later, helped Newton to formulate his famed theory of gravity.

9.3 As They Said of Her

The philosopher and physicist Pierre Gassendi, in writing the biography of Tycho Brahe, said that his sister Sophie was equipped with exceptional knowledge in mathematics and astronomy.

In 1594 Tycho wrote his great Latin poem *"Urania Titani"* in the form of letters from Urania, the muse of astronomy, personified by Sophie, to Titan, represented by Erik, her future second husband. We report here an extract in which the author leaves us a long and affectionate personality description of the different interests of his sister:

> I have a sister named Sophie, who was widowed six years ago. Her husband was a good, noble man. She was still quite a young woman at that time, and had the comfort of one child, a boy. As is so often the case, she had a lot of worries during her years of widowhood, and she began to search for ways to pass the time, that as far as it was possible, could cheer her up now and again. At Eriksholm (her house in Scania, which is built like a fortress) she designed a wonderfully beautiful garden, which is unparalleled in these northern parts of the world. She put enormous effort and inexhaustible work into this project, and other undertakings that I shall not mention, and she laid out the gardens in accordance with the foremost rules, with a well-planned arrangement both of a series of different kinds of trees and garden plants, as well as other suitable details – all in a place where nothing of this kind had existed before. However, when this work was completed she still did not quite free from the weight of her troubles and she sought diversion in chemistry, with the intention of preparing certain spagyric medicines. This was also work that she carried out with no little success. Soon she was not just supplying these preparations to her friends and to the upper classes, where there was a need, but also free of charge to the poor – and thereby she provided them all with great assistance. But she found that not even this road led to the fulfilment of her intellectual ambitions, which continuously aimed further and higher, and finally she devoted great energy to astrological predictions based on birth horoscopes. It might have been her bright talent and one or another Genius that continually prompted her to aim higher and higher, or that her gender in itself predisposed her for curiosity for knowledge of the future, maybe also mixed with a certain amount of superstition. While I myself had given her instruction and guidance within these first areas when she asked for it (admittedly more in chemistry than horticulture, which she herself understood sufficiently well), I strongly advised her to stay away from astrological speculations, since I felt that she should not devote herself to subjects that are too abstract and complicated for the female mind. But she, who has a strong mind and so much self-confidence that she is equal to any man in spiritual matters, ignored my advice and threw herself with increased eagerness into her studies, and taught herself the basics of astrology in just a short time, partly from Latin writers, that she had translated to Danish at her own expense, also partly from German writers in the subject (she has excellent knowledge of German). When I saw the clear signs of this, I stopped working against it, and merely advised her to exercise prudence in her further studies. (Tycho Brahe, excerpts from "Urania Titani")

Few Scanian women have a more deserved reputation than Sophie Brahe, Tycho's sister. She stands for something new in the Nordic woman's history"; wrote the historian Lauritz Weibull more than a century ago. "She is the first and one of the most reliable representatives in the Nordic countries of the Renaissance ideal of womanhood, governed by a fervent devotion to science and art [...].

9.4   Curious Facts

Sophie was born in 1556, but other sources indicate the date of August 24, 1559 instead, in Knutstorp, located in the then-called territory of Scania, which is part of Sweden now, but a Danish possession at the time. Her family belonged to the gentry; her father was Governor of the castle of Helsingborg, which is just in front of the castle of Elsinore, renowned for the story of Shakespeare's Hamlet.

In 1579 Sophie married a wealthy and noble gentleman 16 years older than her, whose name was Otto Thott of Eriksholm, the current Trolleholm and the portrait was painted shortly after the death of her first husband.

They lived a lavish life in his possession, and on May 27, 1580 was born their only son, Tage Ottesen who became a State Councilor, like Sophie's father. Tage grew up loving his mother and later he was one of her few supporters; he died in 1658.

Her domestic engagements did not deter her from the astronomical work at the observatory, and she often visited her brother on the island of Hven, remaining at Uraniborg for several weeks in the occasion of the visit of distinguished guests such as Queen Sofia, in August 1586. After 11 years of marriage, on March 23rd 1588 she became widowed with full responsibility for her son. She took care of his education, also becoming the administrator of the family properties. In order to keep her mind busy, she devoted herself also to the study of astronomy, astrology, chemistry, alchemy, medicine, botany and ornamental gardening. At the time there were no clear boundaries between what we today call science and pseudoscience.

All Tycho had to do in exchange for the island of Hven, was to deliver astronomical almanacs and horoscopes to the court, but did not himself really believe in astrology, in contrast to his sister Sophie.

While Sophie was working with Tycho on his island observatory, she met Erik Lange who visited her brother frequently to discuss his own work in alchemy. They became engaged and Tycho was the only member of the family who supported their relationship.

Years later she married again, but still in contrast with her family, as it is testified also by a letter of Sophie to her sister Margrethe Brahe, where she expresses anger with her family for not accepting her science studies and for depriving her of money owed to her. Probably in consequence of these fights, their economic conditions had to be all but wealthy, since in the same letter she describes having to wear stockings with holes in them for her wedding and returning her husband's wedding clothes to a pawn shop after the wedding because they could not afford to keep them. It is possible, however, that the family was worried with some reason, as this time the marriage was not as happy as she could hope. Her second husband, indeed, died in Prague in misery in 1613, after having squandered their wealth for his stubborn dedication to alchemical studies.

After the death of the second husband, Sophie moved back to Denmark, where she died in 1643 and was buried with the Thott family.

Chapter 10
Maria Cunitz (1610–1664)

"Silesian Pallas" or "The second Hypatia" (Abbé Halma)

Maria Cunitz (Cunitia) was born in 1610 in Silesia, a historical region of Central Europe now belonging almost entirely to Poland. Her home town was Schweidnitz, but some more recent sources indicate the city of Wohlau instead, whose current name is Wolow. She was the eldest daughter of Heinrich Cunitz, an open-minded and wealthy doctor and landowner of that region, and Maria Scholtz of Liegnitz, daughter of the German scientist and mathematician Anton von Scholtz, who was counselor of the Duke Joachim Frederick of Liegnitz. Whatever the actual home-town of Maria, it is attested that eventually the family moved to Schweidnitz, today Świdnica, in Poland.

© Springer International Publishing Switzerland 2016

G. Bernardi, *The Unforgotten Sisters*, Springer Praxis Books in Popular Astronomy,
DOI 10.1007/978-3-319-26127-0_10

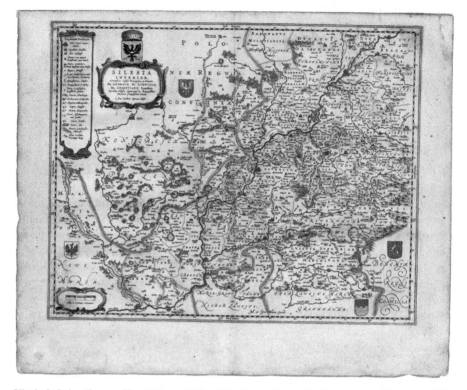

Silesia Inferior (Lower Silesia) in a 1645 publication of Jonas Scultetus from "Theatrum Orbis Terrarum, sive Atlas Novus in quo Tabulæ et Descriptiones Omnium Regionum, Editæ a Guiljel et Ioanne Blaeu" (Theater of the World, or a New Atlas of Maps and Representations of All Regions, Edited by Willem and Joan Blaeu), 1645

As usual for those times, Maria was denied access to university in any form, and she received a broad education by her father on many subjects like languages, painting, music, poetry, mathematics, medicine and history. She married twice: the first time at a very young age, in 1623, with attorney David von Gerstmann who died in 1626. The second time in 1630 with Elias von Löwen or Elie de Loewen, a physician at Pitschen and amateur astronomer as well. He was much older than her, but played a fundamental role in Maria's education, encouraging her to pursue astronomical studies even since before their marriage. This is testified, for example, by records of their observations of the planet Venus made together in December 14, 1627 and of the planet Jupiter in April 1628. They also had three children: Elias Theodor, Anton Heinrich and Franz Ludwig. In 1661 Maria became a widow and then she died herself at Pitschen, on August 22 or 24, 1664 at just 54 years.

10.1 Works

The work for which Maria Cunitz is universally remembered as a scientist is *Urania Propitia*, whose complete title was:

- *Cunitz, Maria. Urania Propitia Sive Tabulae Astronomicae Mire Faciles, Vim Hypothesium Physicarum A Kepplero Proditarum Complexae; Facillimo Calculandi Compendio, Sine Ulla Logarithmorum Mentione Phenomenis Satisfacietes; Quarum usum pro tempore praesente, exacto et futuro communicat Maria Cunitia. Das ist: Newe und Langgewünschete, leichte Astronomische Tabelln, ect.* [Introduction by Elias von Löwen] Oels, Silesia, 1650.

As it can be understood from the title, this work contains, in addition to the ephemeris tables and the description of algorithms for their calculation which were simpler with respect to the original ones from Kepler, general considerations about astronomy and its theoretical foundations introduced with a plain and accessible language.

The importance of this work can be better understood if it is placed in the context of contemporary astronomical research. Maria's astronomical studies, in fact, were quite advanced for the epoch, in fact she was aware of Kepler's research results that supported the Copernican heliocentric system and the ellipticity of planetary orbits, two of the most innovative scientific topics of the seventeenth century. This however did not remain a mere passive topic of her impressive study career. Rather she became involved in active research, even if limited financial resources did not allow her to buy the instruments needed to become competitive in observational astronomy. The Tabulae Rudolphinae are the calculations of the planetary ephemerides performed by Kepler, using Tycho Brahe's observing data, and published in 1628. They were the basis for the research of many other scientists and Maria Cunitz took part in this endeavor by preparing her own version of the planetary ephemeris. Her work distinguished among the others for her original attempt to obtain comparable results with considerably simpler algorithms which required much less calculations than the ones of Kepler. This was a potential breakthrough in an epoch in which any computation had to be done manually: simpler and shorter operations would in fact not only mean faster calculations, but were also less prone to the inevitable computing errors that plagued any table of the ephemerides of that time.

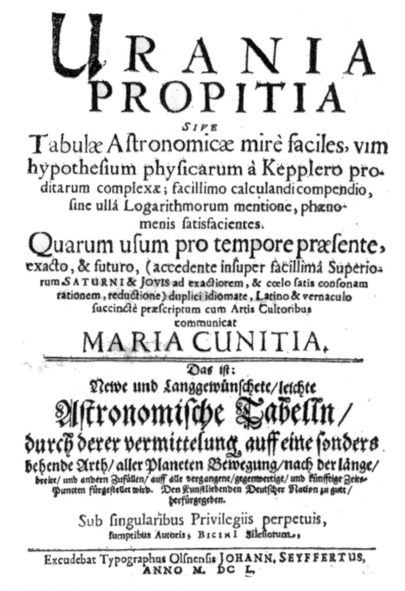

Frontispiece of Urania Propitia by Maria Cunitz

   The outcome of her work was subject to both praise and criticism. She corrected several errors Kepler found in the tables of its boards and simplified his work, but by omitting in her formulas the contributions of some small coefficients, she introduced a number of new errors. It has to be said, however, that the accuracy of her ephemerides were in many cases better than those of Kepler, and that none of the contemporary tables from other scientists were void of computational errors.

10.2 Curios Facts

Maria Cunitz held an epistolary correspondence with other contemporary scientists, even if, because of the conventions of the time, the letters were always addressed to her husband. Among them we can cite the famous astronomer Johannes Hevelius, which we will discuss later, but also the French astronomer Ismael Boulliau or Pierre Desnoyers, secretary of the Queen of Poland and Pierre Gassandi, previously mentioned in the chapter about Sophie Brahe.

The Thirty Years War raged from 1618 to 1648, but Maria Cuniz and her family took refuge in the Cistercians convent of Olobok, managing to carry out her research and finally publishing her results in 1650 in the above cited work *Urania Propitia*. Printed at her own expense, and dedicated to Emperor Ferdinand III, the text is written both in Latin and in German for a greater spread, and today is also credited for her contribution to the development of scientific German language.

It is interesting to notice that in the preface the author herself reassures the readers on their own competence in the matter, and in the next edition the book contains also a declaration of her husband who confirms that his wife is the only author of this work.

Since it was privately published, there is no doubt that a small number of copies were printed, so it is little surprise that it is considered a rare text. Actually, as of today it seems that there are nine copies in the world: one of these, in Europe, is located at the Library of the Astronomical Observatory of Paris, where it can be seen in the afternoon. Another, in the United States, was recently purchased by the Library of the University of Florida:

> The book that has been chosen as the four-millionth volume for the University of Florida Libraries is 'Urania Propitia' by Maria Cunitz. The book examines the theory and art of astronomy, as well as presents her calculations, and a guide to astronomy for nonscientists. According to Cunitz, there were four components to astronomy: carefully recorded observations, the construction of astronomical instruments, theory, and the calculations or tables of predictions. The book is very rare - one of nine copies in existence - and is an important addition to the libraries because it celebrates the university's commitments to women's studies, history of science, astronomy, and the printed word as the prime means of communication for more than five hundred years.

In 1648, after the Thirty Years War, the couple returned to their home at Pitschen, in Silesia, but on the evening of May 25, 1656, at 10 p.m., a large fire destroyed most of the homes in the city. Maria's house was completely destroyed and in such a dramatic event were lost books, letters, over 200 records of her astronomical observations, and instrumentation, both hers and the medical ones of her husband.

Her reputation as a scientist survived her death. Asteroid 12624 was named Mariacunitia after her, and the crater Cunitz on the planet Venus takes her name as well.

Other versions of his last name are Cunicia, Cunitzin, Kunic, Cunitiae, Kunicia or Kunicka.

10.3   As They Said of Her

In 1727 in his book *Educated Silesian Women and Female Poets*, Johan Kaspar Elberti wrote that

> (Maria) Cunicia or Cunitzin was the daughter of the famous Henrici Cunitii. She was a well-educated woman, like a queen among the Silesian womanhood. She was able to converse in 7 languages, German, Italian, French, Polish, Latin, Greek and Hebrew, was an experienced musician and an accomplished painter. She was a dedicated astrologist and especially enjoyed astronomical problems.

But Elberti was more critical when he gave other details of Maria's life:

> [Cunitz] was so deeply engaged in astronomical speculation that she neglected her household. The daylight hours she spent, for the most part, in bed (concerning which all manner of ridiculous events have been reported) because she had tired herself from watching the stars at night.

In his translation of the *Commentaire de Théon Alexandrie*, Abbé Halma sketches the life of Maria and calls her the "*Second Hypatia*".

Maria was a very erudite woman and was also remembered by the famous Italian poet and writer Giacomo Leopardi in his work "*History of Astronomy*".

The origin of Maria Cunitz's dubbing as "Silesian Pallas" comes from J.B. Delambre, who first used it in his study of the history of astronomy, where he also compared her to Hypatia.

Chapter 11
Elisabetha Catherina Koopman Hevelius (1647–1693)

Mother of lunar maps

Another astronomer, always referred to in the letters of Carolina Herchel, is the Polish Elisabetha Catherina Koopman Hevelius who since her childhood dabbled in astronomy. She was born in 1647 in Danzing, and was baptized in January 17. Her mother was Joanna Mennings or Menninx (1602–1679) and her father Nicholas Kooperman or Cooperman (1601–1672) a prosperous local merchant. The two had married in Amsterdam in 1633, and after having transferred from Amsterdam to Hamburg, already in 1636 they moved again to Danzig, a largely German speaking city but a part of Poland at the time.

© Springer International Publishing Switzerland 2016
G. Bernardi, *The Unforgotten Sisters*, Springer Praxis Books in Popular Astronomy,
DOI 10.1007/978-3-319-26127-0_11

Hevelius and his wife Elisabetha making observations in their private observatory

At only 16 years old, in 1663, she became the second wife of Johannes Hevelius, a wealthy merchant of Danzig 36 years her senior. They fortunately shared the same passion for astronomy, which made her marriage a happy one despite the large age difference. She herself became responsible for their private observatory, built on the three roofs of neighboring houses owned by her husband, and served as assistant to the many astronomers who visited him. Elisabetha and Johannes had a son, who died early, and three daughters, the eldest of whom was named Catherina Elisabetha, after her mother. All of them later married and had children of their own. Elisabetha Hevelius worked for more than 10 years with her husband, trying to improve the tables of planetary orbits of Kepler and to fill out a star catalog, until a fire, on September 27, 1679, put this lifelong work at serious stake. The fire had been caused by a candle left burning in the stable by Hevelius' coachman, and it destroyed not only the fine brass instruments of the observatory, their astronomical data and most of the copies of the work *Machinae celestae*, but also their private library and typography. Despite such a severe blow and his old age (he was 68 at the time) the astronomer resolved to pursue his decades-long scientific enterprise and, thanks also to some powerful patronages, among which there were also the kings of France Louis XIV and that of Poland John III, the observatory was eventually rebuilt and the observations started again. It is argued that his resilience was at least in part fueled by the miraculous salvation of one of his manuscripts, a small leather-bound notebook which contained the results of thousands of calculations of the positions of the stars made over decades of patient observation. It was rediscovered three centuries later, and in 1971 was bought by the Utah's Brigham Young University, becoming the one-millionth acquisition of its library. In celebration of this event, the university published a volume entitled *"Johannes Hevelius and His Catalog of Stars"* which reported the life and legacy of Hevelius, and the 300-year odyssey of his fixed-star catalog.

11.1 Works

Johannes Hevelius had fallen into disrepair after the destruction of his work and instruments, and after her husband's death on January 28, 1687, the exact date of his 76th birthday, Elisabetha continued their research alone, publishing the results of their observations in a work entitled *Prodromus Astronomiae* in 1690.

Johannes Hevelius, Prodromus Astronomia, volume III: Firmamentum Sobiescianum, sive Uranographia, table 55: Northern Hemisphere, 1690

Strictly speaking, this work is just the first of three publications contained in a single book. The other two are the *Catalogus Stellarum*, and an atlas of constellations named *Firmamentum Sobiescianum sive Uranographia*.

In particular, the *Prodromus* outlines the methodology and technology used in creating the star catalogue and it provides examples of the use of sextant and quadrant by Johannes in tandem with known positions of the sun, in calculating each stars' longitude and latitude. The *Catalogus* is the most extensive stellar catalog obtained without the use of the telescope still in existence, and contained the exact location of a good 1564 stars. Finally, the Firmamentum Sobiescianum takes his name from the dedication to the King of Poland John III Sobieski and comprises 73 constellations.

As specified in the complete title, the *Catalogus Stellarum* merges the newly computed positions from their original observations with those of other catalogs from previous astronomers, specifically Tycho Brahe, the Landgrave (Prince) William IV of Hassia, Giambattista Riccioli, Ulugh Begh, and Ptolemy. These are reported in tabular form, with several columns recording for each star (ordered alphabetically according to the constellation name) its magnitude as estimated by themselves and by Tycho, the ecliptic longitude and latitude of all the above mentioned catalogs, and their right ascension and declination computed using spherical trigonometry. Since the catalog of Hevelius included about 600 new stars and a dozen new constellations not present in the other ones (some of which are still named as in this catalog) each constellation is divided into two subgroups: a first one common to all the catalogs and a second with fainter stars available only here which obviously required a narrower table. One of the most notable things of this catalog is Hevelius' choice of doing the observations using nothing more than the astronomer's naked eye. Such a decision might appear weird and not optimal for a period in which telescopes were starting to impose themselves over traditional techniques, thanks also to technological improvement. However the fact that their measurements were used in the making of celestial globes into the early eighteenth century speaks for itself. Moreover, considering that the couple owned one of the most advanced telescopes of the time suggests that this was a well-considered resolution. It might be that this was connected to his particular acute sight (or of both of these reasons), and the fact that the new stars of the catalog are all at the faintest end of the range visible with the naked eye, with some of them even classified as of seventh magnitude, seem to come in support of this hypothesis. However it might simply be that telescopes were not yet the best choice for precision astrometry. The accuracy of position astronomy when using telescopes, in fact, crucially depends on factors like the absence of optical defects and on their precision maneuverability, rather than on the light-collecting capability, some characteristics which could have not developed enough at that time.

Firmamentum Sobiescianum, while technically part of the *Prodromus Astronomiae* as a well, was likely published also separately and in tighter circulation. Produced with its own cover page and page-numbering system, the atlas consisted of two hemispheres and 54 double-page plates of 73 constellations. Both the northern and southern hemispheres were centered on an ecliptic pole, and most star locations were all based on Johannes' own observations. Those that were not, the southern polar stars, were based on a catalog and map published in 1679 by Edmond Halley. One specific characteristic of the maps of this book is that they are represented as if they were seen from the "outside" of the celestial sphere. This decision was later highly criticized by the Royal Astronomer Flamsteed since they provide a mirrored representation of the sky, that is reversed with respect to what can be seen by direct observation.

Johannes Hevelius, Prodromus Astronomia, volume III: Firmamentum Sobiescianum, sive Uranographia, table 56: Southern Hemisphere, 1690

11.2   Curious Facts

The Astronomical Observatory of Hevelius actually consisted of three houses huddled together between the numbers 33 and 35, of which Hevelius had joined the roofs to build his own astronomical observatory, one of the finest and best equipped in Europe: the Stellaeburgum, or "village of stars".

The three buildings had been joined many years earlier, when Hevelius had married his neighbor, Katharine Rebeschke, 2 years younger than him and who, as he had started his studies of the sky on the roof, managed the family brewery in the floors below.

Despite his convinced preference for observations made with the naked eye, in his observatory Hevelius built the most powerful telescope of the time, which didn't have a tube and was 150 ft (46 m) long, for the study of the surfaces of the Moon and planets.

On the same roof, the little Elisabetha set foot for the first time when she was still a child, immediately kidnapped by the wonders that this patient and so self-confident man, showed her:

When you have the right age - the promises astronomer - I'll show you all the wonders of the sky.

She took him at his word. In 1662, a few months after the death of the first Mrs. Hevelius, Elisabetha presented herself: she was 15 years old, now grown up (at least for that epoch) and was ready to collect the promise. Hevelius was 52 years old when the young and devoted Elisabetha became his second wife.

The help of Elisabetha was not negligible: she could do calculations and handle the complex instrumentation on the roof. Also, she knew Latin even better than her husband, helping him to keep in contact with other European astronomers. Actually Hevelius joined the Royal Society of London in 1664.

The most famous image of the Hevelius couple is from an engraving of *Machinae Celestae* which depicts Elizabeth and her husband in the act of making an observation with a large brass sextant.

The first star maps of a planned series which eventually flowed into the Firmamentum Sobiescianum was published, with Elisabeth's help, in 1673, and when, after the publication of her "Magnus opus" in 1690, Elisabetha finally brought to an end the lifelong quest for their stellar catalog, she continued to guard the manuscripts until her death, in1693. A complete set of their published works were left to each of their three daughters, but Catharina received a special edition of that book, originally prepared as a gift for the king of France Louis XIV. The reason can be probably explained as a sign of gratitude to the person who, at the age of 13, presumably saved the manuscript of the fixed-star catalog from the disastrous fire of 1679. The story of this document is indeed quite singular and of particular interest, since, after the first rescue it escaped fortuitously from destruction two more times up to our days. Once Catharina married, her husband sold most of Hevelius's prized books to a museum in Russia, but the manuscript of the star catalog that had survived the fire was overlooked. Ironically, the greedy son-in-law didn't judge the original copy of Hevelius' main work valuable enough to sell. Then it survived the destructions of at least two more wars: one of 1734 when, during a siege of Danzig, a bomb destroyed almost everything of the building where it was conserved, and at the end of World War II when once again its existence was put at stake by a fire caused by the fighting.

Elisabetha Hevelius died on December 23, 1693, at the age of 46, and was buried in the same tomb as her husband; her's life was dramatized in 2006 in the novel *"The Star Huntress"*.

11.3   As They Said of Her

Elisabeth Catherina Koopmann Hevelius is considered one of the first women astronomers, and is also called the *"mother of lunar maps"*. Her passion for astronomy is witnessed by a sentence of hers, as quoted by a German biography: *"To remain and gaze here always, to be allowed to explore and proclaim with you the wonder of the heavens; that would make me perfectly happy!"* It has to be told, however, that, since it was told to Hevelius one night before their marriage, while looking through his telescope, it was essentially a marriage proposal, which he gladly accepted.

On February 3, 1663, when Johannes was 52 and Elisabeth was 17, they were wedded in Danzig at St. Catherine's Church. Despite their big age difference, there are many clues that they had a happy marriage. Indeed, such unions were not so uncommon at the time, and moreover they had the advantage, from the point of view of the women, to grant access to creative and intellectual pursuits through a kind of conjugal apprenticeship, which otherwise would have been impossible in a society where female education was almost completely disregarded.

Another biography cites Elisabeth's most frequent words of encouragement to Hevelius: *"Nothing is sweeter than to know everything, and enthusiasm for all good arts brings, sometime or other, excellent rewards."*

From the writings of the mathematicians Johann Bernoulli III we know that Elisabetha: *"...was taken ill with small-pox and was badly marked by it. Her husband, who had never had this disease, never left her sick-bed and nursed her faithfully..."*

Elisabeth worked alongside Hevelius in completing the star catalog that had become the holy grail of his scientific career and his highest hope for a lasting legacy. Hevelius always spoke highly of Elisabeth's scientific skills. One telling example of his consideration can be found in a letter to the king of France, which included an engraving of the duo making an observation together. Remembering the dreadful accident which destroyed their home and the observatory, he wrote: *"On the unhappy evening before the fire I felt deeply troubled by unaccustomed fears. To lift my spirits, I persuaded my young wife, the faithful assistant for my nightly observations, to spend the night in our country retreat outside the walls of the city ..."*

The British astronomer Edmond Halley had visited Hevelius and Elisabetha during the summer of 1679, shortly before their home and observatory were destroyed by fire. This shocking news apparently reached Halley after his departure from his hosts, and it seems that a little misunderstanding happened in the communications, since he sent a dress not directly to Elisabetha, who had requested it, but to the secretary of the city of Danzig, together with the following letter:

> I quite realise that [Hevelius's] heartbroken spouse must be wearing sad-coloured apparel, yet for several reasons I have thought well to send the gown procured for her . . . because I am not yet certain her husband is dead, in which case I judge nothing would be more unwelcome than delay . . . since it is of silk and of the newest fashion, I am confident it will highly please Mme Hevelius, if only it should be granted to her to wear it . . .

The mathematician Francois Arago, after her death, wrote of her character: "*A complimentary remark was always made about Madam Hevelius, who was the first woman, to my knowledge, who was not frightened to face the fatigue of making astronomical observations and calculations*".

The asteroid n. 12625 Koopman was named in his honor, as the crater Corpman located on the planet Venus.

Chapter 12
Jeanne Dumée (1660–1706)

[...]because there is no difference between the brain of a man and that of a woman.

In 1660 Jeanne Dumée was born in a middle-class family of Paris. Since young Jeanne developed a very strong passion for science, and her lively intelligence was stimulated by the activity of the newly founded Royal Academy of Sciences. This institution had been established by King Louis XIV, and was later joined by a Royal Library and the renowned Astronomical Observatory, immediately attracting great scientists and astronomers. Among them we can remember the Dutch Physicist, Mathematician and Astronomer Christiaan Huygens, who became the first director of the Academy of Science, and the Italian astronomer Gian Domenico Cassini, the first director of the Astronomical Observatory and progenitor of a celebrated family of astronomers.

© Springer International Publishing Switzerland 2016
G. Bernardi, *The Unforgotten Sisters*, Springer Praxis Books in Popular Astronomy,
DOI 10.1007/978-3-319-26127-0_12

Giovanni Domenico Cassini, first director of the Paris Observatory appearing in the background

She got an early marriage with a professional soldier, who died at war in Germany. At the age of 17, widow of a relatively high-ranking officer, she was left with some substance, which allowed Jeanne a certain degree of independence in pursuing her thirst for knowledge, and in particular in Astronomy.

12.1   Works

Jeanne fit out her attic as a private astronomical observatory, provided with a telescope that she equipped with the latest inventions like a micrometer, improved ocular lenses with long focal lengths. This was the natural endpoint of a serious educational path she had started by reading everything she could find on the subject. In Paris, at her time, there existed 23 private observatories, and she managed to visit all of them. She was also able to see the Royal Observatory that Louis XIV inaugurated in 1683 before its completion.

She became exposed since the very beginning to the modern ideas of her time like Copernicanism and, thanks to the vivacious scientific activity of her city, Jeanne could become aware of the latest advances of astronomy, like, e.g., the breakthroughs of Gian Domenico Cassini about Jupiter, Mars and Saturn, or the big map of the Moon presented in 1679 at the Academy of Sciences.

Her deep knowledge of the subject, and her astronomical observations gave her a great authority within the cultural "salons" of the epoch, where Jeanne could shine among her peers thanks to exposures of the new thesis.

A gathering of distinguished guests in the drawing-room of French hostess Marie-Thérèse Rodet Geoffrin (1699–1777) who is seated on the right. Such kind of events constituted a common practice of the "Salons" which became common in the high society between the seventeenth and eighteenth century

Soon she became well known and held popular conferences for an enlightened public. The success of her exhibitions pushed her to build two astronomical spheres for an easier and more enjoyable exposition. The first was a traditional armillary sphere representing the Earth at the center with the Sun, the Moon and the planets around; the second sphere was instead based on the new Copernican cosmography, showing the Sun at the center and the planets revolving around it.

Her conferences were so successful that she was persuaded to write, in 1680, a work in which she explained in detail the three motions attributed to the Earth and provided arguments in support and against the Copernican system. It was entitled

- *Entretiens sur l'opinion de Copernic touchant à la mobilité de la Terre (A Discussion of the Opinion of Copernicus Concerning the Mobility of the Earth)*, dedicated to Chancellor de Boucherant.

What I claim here on the ideas of Copernicus has not the purpose of establishing, - as she wrote prudently - and much less of supporting them, but only of showing the reasons by which the Copernicans defend themselves, and also to please some learned man, that honoured me with his visit, who saw a sphere that I had prepared after these opinions.

They forced me to popularize the reasons of these theses, and they urged to write them down.

The *Journal des sçavants* announced the publication of her book and praised the author for having clearly explained "the three movements of the Earth;" it is not certain however that the work was published, since only the National Library of Paris has a manuscript.

12.2   Curious Facts

Although the importance of Jeanne Dumée for her contribution to spreading the Copernican system is well established, it is not known what pushed her to the study of astronomical science. It is little doubt, however, that if her husband had not died prematurely, he would have surely deplored the passion of his wife for the observation of the sky as a crazed habit. Another objection that he could have raised was about her theses, which we would now define feminists. Jeanne in fact was not content to be learned, but she also wished that other women could follow her example. Wrongly, she says, they believe themselves to be unfit to the study of astronomy:

> On dira peut-être que c'est un ouvrage trop délicat aux personnes de mon sexe ; je demeure d'accord que je me suis laissé touchée à l'ambition de travailler sur des matières auxquelles les dames de mon temps n'ont encore point pensé et même afin de leur faire connaître qu'elles ne sont pas incapables de l'étude, si elles s'en voulaient donner la peine, puisqu'entre le cerveau d'une femme et celui d'un homme il n'y a aucune différence. Je souhaiterais que mon livre leur put donner quelque émulation.

> They may say that this is a too good work for person of my sex; I agree that I let myself to be taken by the ambition of working on subjects that the women of my time have not considered yet, and even to make them aware that they are not disinclined to such studies, if they want to take the effort, because there is no difference between the brain of a man and that of a woman. I wish that my book can give them some example.

Unfortunately, it seems that this work, at least in the immediate, did not contribute a lot to the realization of her wish. It apparently did not enjoy a large spread if, as we know, people like the French astronomer Jérôme de Lalande, who sought a printed copy, was not able to find any. However, as we already mentioned above, it is possible that the manuscript was never published even if the German scientist Johann Friedrich Weidler in his *Historia Astronomiae sive de Ortu et Progressu Astronomiae* (History of Astronomy, or on the rise and progress of Astronomy) published in 1741, mentions it as an actual publication.

The manuscript of the *Entretiens sur l'opinion de Copernic touchant à la mobilité de la Terre*, is now located at the Bibliothèque Nationale of Paris, with catalogue number 19941.

12.3   As They Said of Her and She Said of Herself

Jeanne Dumée is cited among women astronomers by Jérôme de Lalande in his text *"Astronomie des dames"* (Astronomy of the ladies, Paris, Bidault, 1806, p. 6), as well as, in more recent times, by Jean-Pierre Poirier in his book *"Histoire des femmes de sciences en France"*, 2002.

Curious, independent and self-confident, Jeanne wrote already in the seventeenth century: "...*entre le cerveau d'une femme et celui d'un homme il n'y a aucune différence*" or "...*there is no difference between the brain of a man and that of a woman.*"

Part III
Timelines from Maria Margarethe Winkelmann-Kirch to Nicole-Reine Étable de la Brière Lepaute

1718 Halley measures the first proper motions of the stars

1729 James Bradley announces the discovery of aberration of light and the phenomenon of nutation

1751 Diderot and D'Alambert start to publish the Encyclopaedia

1755 the German philosopher Kant exposes the first modern theory of the origin and evolution of the Solar System

1774 Charles Messier publishes his catalog of celestial objects

1776 Declaration of Independence of the United States of America

1781 William Herschel discovers Uranus, 1788 the Italian mathematician Lagrange publishes *Analytical Mechanics*

1787 Lavoisier publishes a treatise of elementary chemistry in which he states the Law of Conservation of matter

1789 French Revolution

1800 William Herschel discovers infrared rays by observing sunlight

Chapter 13
Maria Margarethe Winkelmann-Kirch (1670–1720)

A comet of one's own

Again in the Teutonic lands, we move to Panitsch, near Leipzig, in the German state of Lower Saxony, to meet the first woman who officially discovered a comet: Maria Winkelmann, born on February 25, 1670. Since very young she was educated by her father, a Lutheran pastor, who believed in an equal footing education for both sexes. At the age of 13, she lost both her parents and her education continued under her guardian, uncle Justinus Toellner. Maria Margarethe became interested in astronomy already in a very young age and she had the opportunity to become a student, apprentice and then finally assistant of Christoph Arnold Sommerfeld known as the "peasant astronomer," who discovered a comet in 1683. It was probably in this circle that she had the opportunity to meet her future husband, the astronomer and mathematician Gottfried Kirch (1639–1710) who became one of the most famous German astronomers of the time. He published a long series of calendars and ephemeris and discovered a comet in 1680, which became the first comet in history discovered with a telescope. Later, in 1686, he also discovered the X-Cygni variable star, the third known at the time. He was born during the Thirty Years War and his father Michael Kirch, a tailor, had to flee with his family from his native town of Guben. Kirch lived a quite restless childhood, and he probably did not get a degree, but he had good academic contacts, as for example Erhard Weigel, professor of mathematics at the University of Jena from 1653 to 1699, who recommended him to famous astronomer Johann Hevelius. Thanks to this recommendation, he could work for a short time, in 1674, in Danzig in Hevelius' well-appointed private observatory, probably meeting also his second wife, the astronomer Elisabethe Koopman. Before reaching tenure Kirch maintained himself with the creation of calendars and as a teacher, living in various places like Langgrun, where in 1667 he married his first wife Maria Lang who died in 1690 after having given birth to seven sons and one daughter. Although the Winkelmann's family was originally against the union, because of the 30 years of difference, and since they wished Maria Margarethe married to a Lutheran pastor, in May 8, 1692 she married Kirch, maybe following her astronomical interest. They had seven children, five daughters and two sons, who were all educated from the earliest age to the family

© Springer International Publishing Switzerland 2016
G. Bernardi, *The Unforgotten Sisters*, Springer Praxis Books in Popular Astronomy,
DOI 10.1007/978-3-319-26127-0_13

business of astronomy, but apart from the firstborn Christfried, Christine and
Margarethe, there is no information about the others. They initially lived in Saxony,
in Leipzing and Guben, for a few years before moving in 1700 in Berlin where
Kirch accepted tenure.

13.1   The Kalenderpatent

Settled in Berlin, Kirch directed the studies of his wife as he had done with his three
sisters, who even after his tenure as full astronomer, on May 18, 1700, maintained
themselves mainly by producing calendars, but also almanacs, books of observations and calculations.

A 1824 reproduction of the Old Berlin Astronomical Observatory officially opened in 1710

The production of calendars deserves a more detailed explanation. This appoint-
ment, called *Kalenderpatent*, was specifically created by Frederick III, Elector of
Brandenburg, in his edict of 10 May 1700. This edict followed the decision of the
German protestant states to introduce, since the year 1700, a new and improved

calendar, identical in practice to the Catholic Gregorian except for the date of Easter, that would have been calculated by qualified astronomers. This edict thus introduced a monopoly in the Electorate of Brandenburg, and later by Prussia, for the realization of this calendar and the imposition of a "calendar tax" whose proceeds were used to pay the astronomers and the other members the Academy of Sciences in Berlin. The academy was founded in the same year, on July 11, and Frederick III also promised the creation of an observatory in Berlin, then inaugurated on January 19, 1711. Kirch produced in 1700 the first calendar of the series: *"Chur Brandenburgischer Verbesserter Calendar Auff das Jahr Christi 1701"* and in this endeavor he was assisted by his wife Maria Margarethe. Although the observing conditions in Berlin were not the most favorable ones, the Astronomical Observatory continued in its implementation, while Kirch's family pursued its duties by observing with small transportable instruments which, at first, were placed on the roofs of their house, and after 1708 on the tower of the still unfinished Observatory. From here Maria Margarethe alternated for years with her husband in the observing nights at the telescope, but beyond that she also reserved for herself the job of computing the ephemerides. It was customary at the time to incorporate all of the findings under the authorship of her husband, so it is not surprising to know that Mary's discovery of a comet of 1702 was made public with the name of her spouse and that only after a few years it was officially assigned to her.

13.2 Discoveries and Works

Just before the morning of March 21, 1702 Maria Margarethe, during an observing night, discovered comet C/1702 H1 which was then called the "Comet of 1702". In the words of her husband: *"Early in the morning the sky was clear and starry. Some nights before I had observed a variable star, and my wife wanted to find to see it for herself. In doing so she found a comet in the sky. At which time she woke me and I found that it was indeed a comet … was surprised that I had not seen it the night before."*

Kirch later confirmed that the discovery was actually to be attributed to his wife.

Among her works, published under the name of Maria Margarethe, remain some treaties concerning the observation of the aurora borealis, or northern polar lights, of 1707, the brochure *Von der Conjunction der Sonne des Saturni und der Venus* (on the conjunction of the Sun to Saturn and Venus) of 1709, and the forecast of the conjunction of Jupiter and Saturn in 1712, which included the astronomical and astrological observations, as was the custom, especially during the conjunctions of planets, although Maria Margarethe tried to distance herself from astrologers.

All the other works, mainly concerning the calculations for calendars and their observations, were published under the name of her husband or of their son.

13.3   Curiosity

When Kirch died on July 25, 1710, Maria Margarethe continued her observations in spite of various obstacles. Indeed, she was at first denied the appointment of the preparation of the calendar because, even if the work had always been actually done by herself, the Academy of Sciences had officially appointed her husband for this task. Wishing to avoid the precedent of an official presence of a woman at a public institution, the Academy, despite the open support of the president—the physicist-philosopher Gottfried Leibniz—rejected her application, even with a minor task, which nonetheless was necessary in order to continue her calendars' work. The secretary of the Academy, Johann Jablonski, wrote: *"Already during her husband's life the society held up for public ridicule as the calendar was prepared by a woman. If now she was allowed to continue with such appointment, the astonishment would be even greater."* The successor of her husband to the Direction of the Observatory and the compilation of the calendars became, in 1711, Johann Heinrich Hoffmann, one of his students.

After the final rejection of 1712, Maria Margarethe expressed her disappointment in this way: *"Now I go through a severe desert, and water is scarce because . . . . . . the taste is bitter."* In the same period she wrote in the foreword of one publication of hers that a woman could become *"as skilled as a man at observing and understanding the skies."*

In October 1712 Maria Margarethe, together with her sons, was admitted as an astronomer at the Berlin private observatory of Baron Bernhard Friedrich von Krosigk, where she worked as a teacher.

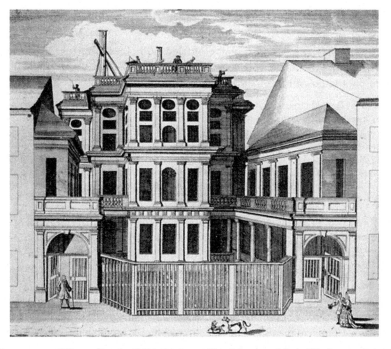

The Observatory of Baron Bernhard Friedrich von Krosigk, were Maria Kirch served as teacher after her husband death, in an engraving by G.P. Busch

She trained her daughter Christine and her son Christfried as assistants, publishing ephemeris and continuing her work on the calculation of calendars for the cities of Wroclaw (the German Breslau), Nuremberg and Dresden as well as for Hungary, until her death. When the Baron died, she moved to Danzig, to rearrange and use the observatory of the famous Johannes Hevelius. In 1716 she was proposed to become astronomer at the court of the Tsar of Russia Peter the Great, but she declined the offer because the son Christfried had become director of the Berlin Observatory, after the death of Hoffmann, and she was implicitly required to stay at his side as an assistant, as much precious as invisible when guests were visiting the observatory. She was eventually removed from the observatory by the Academy of Sciences, because she refused to stay in the shadows, and she died 3 years later, on December 29, 1720, at 50. Her daughter Christine assisted the brother in observations and astronomical calculations and for many years she was entrusted with the calculation of the calendar for Silesia.

13.4   As They Said for Her

The Academy of Science's president, Gottfried Wilhelm Leibniz, in 1709 presented Maria Winkelmann to the Prussian court, where she explained her sighting of sunspots. In a letter of introduction Leibniz wrote:

> There is [in Berlin] a most learned woman who could pass as a rarity. Her achievement is not in literature or rhetoric but in the most profound doctrines of astronomy.... I do not believe that this woman easily finds her equal in the science in which she excels.... She favors the Copernican system (the idea that the sun is at rest) like all the learned astronomers of our time. And it is a pleasure to hear her defend that system through the Holy Scripture in which she is also very learned. She observes with the best observers, she knows how to handle marvelously the quadrant and the telescope. (Letter to Sophie Charlotte, January 1709)

Alphonse des Vignoles, vice president of the Academy in Berlin, wrote in 1721: *"If one considers the reputations of Madame Kirch [Winkelmann] and Mlle Cunitz, one must admit that there is no branch of science … in which women are not capable of achievement, and that in astronomy, in particular, Germany takes the prize above all other states in Europe."*

And he also said in her eulogy: *"Madame Kirch prepared horoscopes at the request of her friends, but always against her will and in order not to be unkind to her patrons."* Thus suggesting that she simply had a financial interest in pursuing this kind of non-scientific activity.

The historian of science Lettie Multhauf, in her Dictionary of Scientific Biography, wrote: *"She went to work in Krosigk's well-equipped observatory in 1712, and upon [Krosigk's] death in 1714 moved to Danzig. Peter the Great wanted her to come to Russia, but when her son Christfried, became the astronomer of the Berlin Observatory, she joined him there."*

13.5   Awards

Maria Winkelmann's work was widely celebrated, and in 1711 she received an Academy medal.

The minor planet n. 9815 was named Mariakirch after the astronomer Maria Margarethe Winkelmann Kirch.

Chapter 14
Maddalena (1673–1744) and Teresa (1679–1767) Manfredi

Be less curious about people and more curious about ideas.
(Marie Sklodowska Curie)

Maddalena and Teresa were two sisters of a petty bourgeoisie family of Bologna, Italy. The former was born in 1673 and her sister 6 years later, in 1679. Although their father, Alfonso Manfredi, was a notary, the family was not at ease in supporting all its members, so their mother Anna Maria Fiorini sent three of the four brothers at the University of Bologna for their studies, while the education of the daughters ended up in a convent of tertiary nuns. Maddalena and Teresa learned Astronomy, Mathematics and Latin within the family and the circle of friends who frequented their house. One of the brothers, Eustachio Manfredi (1674–1739), became a public reader of mathematics at the University of Bologna, and from university professor 1699. He was a distinguished scientist and astronomer, and thanks to him the private education of the sisters enjoyed a close contact with the brothers and their writers and scientists friends. Eustachio was an advocate of an eclectic culture and in his own house, together with his friends, he encouraged conversations on different topics such as history, literature, and experimental physics.

© Springer International Publishing Switzerland 2016
G. Bernardi, *The Unforgotten Sisters*, Springer Praxis Books in Popular Astronomy,
DOI 10.1007/978-3-319-26127-0_14

1792

Eustachio Manfredi

Eustachio Manfredi, first director of the observatory in Bologna and brother of Teresa and Maddalena

It was from these, so to speak, "homemade meetings" that Eustachio founded, just 16 years old, the *Accademia degli Inquieti*, literally "Academy of the Restless," devoted to literature and experimental sciences. This cultural circle aimed to connect with the European culture, and in the future it marked a turning point for the study and research of the University of Bologna. Among their distinguished guests of the family one can find the physician Giovanni Battista Morgagni (1682–1771), the poet Pier Jacopo Martello (1665–1727), the poet and naturalist Ferdinand Antonio Ghedini (1684–1768) and the philosopher Francesco Maria Zanotti (1692–1777), the physicist Francesco Algarotti (1712–1765) and many others. A particular bond of commonality linked the Manfredis to the family of the painter and art historian Giampietro Zanotti (1675–1765) and his daughters Maria Teresa and Angiola Anna Maria. With the exception of the youngest brother Emilio, the only one who did not attend University to became a Jesuit, the other brothers had not much different fortunes: both studied medicine, and Gabriele (1681–1761) studied medicine becaming a lecturer of Mathematics and author of famous works on differential calculus, while Heraclitus (1682–1759) after his was first appointed as honorary lecturer of Astronomy and then professor of Hydrometry.

The positions held by the illustrious brothers Manfredi, although relevant in the cultural landscape, were poorly paid. Because of this all but florid financial condition only one of them, Gabriele, got married. These reasons favored the formation of a tightly linked "enlarged family" committed to the creation of a cultural enterprise which could help to improve the common budget. Maddalena and Teresa devoted their physical and mental strength to the administration of the family business, to the scientific collaboration in the works of their brothers, and to the production of literary works for the market of the educated bourgeoisie of Bologna. Eustachio Manfredi became the head of this family, and at the beginning of 1700 they all moved into the palace of earl Luigi Ferdinando Marsili (1658–1730), who, though engaged in military campaigns, was eager to create an academy in Bologna on model of the French Académie Royale des Sciences in Paris, and of the London Royal Society.

Bologna's view with the Astronomical Observatory tower. Antonio Basoli (1826)

Eustachio gave the Count a crucial help for the creation of the Academy of Sciences of Bologna with a decidedly experimental imprint. Meanwhile, in his home had been prepared an astronomical dome and is known from direct witness that the brothers and the sisters got engaged in helping with astronomical calculations up to 1711, when Eustachio was appointed astronomer of the Academy and moved in Via Zamboni in new headquarters. The sisters Maddalena and Teresa followed him even in this event, and while remaining purposely in the shadows of the brother, they gained some fame thanks to literature.

14.1   Works

Maddalena and Teresa never signed their works, either literary or scientific. But their contribution was significant to restore the international fame of the astronomical studies of Bologna.

Their cooperation in the family enterprise consisted in astronomical observations and mathematical calculations of the *Ephemerides motuum Coelestium* (Ephemerides of Celestial motion, 1715) of their brother Eustachio. This work was considered for decades throughout all Europe the most extensive and complete tool for the astronomical characterization of the sites among many manufactured in Europe, thus greatly contributing to the reprise of the local University. At the end of the manuscript of Eustachio Manfredi's *Introductio in Ephemerides*, preserved in Bologna, we read the following note, which is missing in the printed edition: "*I started the ephemeris on December 1712 in Bologna. With many interruptions they continued in the following years with the help of my two sisters Maddalena and Teresa, and of Mr. Giuseppe Nadi, and yet some more from Mr. Cesare Parisij . . . The table of longitudes and latitudes was calculated by my sister Maddalena in 1702 or 1703.*"

The astronomical work of the brothers and sisters Manfredi and of their friends, despite its high scientific level and advanced technology, was a typical "home-made" product of the eighteenth century. The astronomical observations were first made from the house of Vittorio Francesco Stancari (1678–1709) and then in that of the Manfredis, and the instruments, which included a 5-m telescope and a sextant with telescopic sights, were built by themselves. The calculations were the result of a long and tedious job and for the first time they developed to the point that could become accessible also to non-specialists, who could then be employed in their realization.

The compilation and printing of the *Ephemerides Bononienses*, for whose realization were used the so-called "Cassini tables" made in Paris, as mentioned above was made by Eustachio Manfredi with the important help of the sisters Teresa and Maddalena, but it is possible that a third sister, Agnese Manfredi, also collaborated in this work.

Calliope, Urania et Terpsichore, oil on canvas, 1746

In the end they contained everything that could be useful for the astronomical determination of the geographic coordinates of a site. They were also accompanied by an introductory volume entitled *Introductio in Ephemerides* that described in detail their use, along with the various operations that had to be carried out for the set-up of an astronomical station.

- E. Manfredi, *Ephemerides motuum celestium ex anno MDCCXXV in annum MDXXL,* Bologna, 1725.

14.2 Curious Facts

Aside from the astronomical work, Eustachio also had the collaboration of the sisters also to perform literature research for his erudite writing *"Compendiosa Informazione di fatto sopra i confini della comunità ferrarese di Ariano con lo Stato Veneto"* (Concise information on the borders of the community of Ariano, near Ferrara, with the Venetian State, 1735).

The practice of leaving their works unsigned has to be traced back to the use of the epoch, but maybe also to a personal habit of considering their work mainly with respect to a somewhat limited circle, without seeking any fame outside of it. For example, they never signed the translation of their *Bertoldo* and *Chiaqlira*. However it was generally knew that they were the authors, especially among the representatives of the intellectual middle-class of Bologna who loved to invite the two sisters to their parties and events. From this point of view, therefore, there was no need to sign them.

The Manfredi family always cultivated a twofold passion for poetry and science, which was typical of the eighteenth century, so it is not surprising to learn that their members sided their astronomical publications with literary works for erudite as well as ordinary people. Eustachio became a member of the prestigious *Accademia della Crusca* for his Latin verses, while the sisters Maddalena and Teresa will delight themselves with the dialect. Their first work was the translation of stories from the Neapolitan to the idiom of Bologna and their integration with new songs, proverbs, rhymes and allegories: *Bertoldo con Bertoldino and Cacasenno in ottava rima aggiuntavi una traduzione in lingua bolognese* (1740). Shortly after this one, in 1742, they translated other Neapolitan verses in *La Chiaqlira dla Banzola o per dir mìi Fol divers tradutt dal parlar Napulitan in lengua Bulgnesa per rimedi innucent dla sonn, e dla malincunj dedicà al merit singular del nobilissm Dam d'Bulogna...* (1742). These two works had such a great success that 2 years later came the translation of the *Pentamerone* by the seventeenth century writer Giovan Battista Basile, a collection of delightful tales once again in Neapolitan, also known as *Lo cunto de li cunti*—The tale of the tales—which will have a prominent influence on tales authors like Perrault or the Brothers Grimm.

- Translation in the idiom of Bologna of *Bertoldo with Bertoldino and Cacasenno*, Bologna, Lelio by Volpe, 1740.
- *La Chiaqlira dla Banzola o per dir mìi Fol divers tradutt dal parlar Napulitan lengua Bulgnesa per rimedi innucent dla sonn, e dla malincunj dedicà al merit singular del nobilissm Dam d'Bulogna...* (1742)
- Translation in the idiom of Bologna of *Pentamerone*, by Giovan Battista Basile, Bologna, 1742.

Maddalena died at 72 on March 11, 1744, while her sister Teresa ended her life 23 years later, on October 8, 1767.

14.3   As They Said of Them

According to the writer and historian Giovanni Fantuzzi, the Manfredi sisters acquired "*a great knowledge of the tables and astronomical calculations* [...] *[so that the first two tomes Ephemeris you must, if not all, most, however, the diligence and the study of these two calculators*" [Fantuzzi 1786–1789, p. 188].

According to the Italian writer Ilaria Magnani Campanacci, the lives of two sisters Manfredi is characterized by a sense of proportion and concreteness which was balanced between a growing awareness of their own talent and acceptance of a role submitted to the public male figure of their brother: "*It is clear from many indications, which can be extracted either from the data of their eighteenth-century biographers as well as from their rhymes . . . that in the recurring male controversy against the knowledgeable woman they speak openly and lively in favor a woman's right to make use of her mind in spite of those who would like her stupid [. . .]. But they firmly reject as well the possible temptation of attributing them a sententious character [. . .] thus in agreement, even if with an unknown degree of self-irony, with the dominant negative opinion about a woman who makes a undue and harassing show of her knowledge. [. . .] In short, the Manfredi were the practical demonstration of the female genius [. . .] applied to the study, but rather as a complement instead of alternative to the commitments of a woman. Not with a competitive attitude, but in close and supportive collaboration with the more recognized intellectual figures of their academic brothers*" [Magnani Campanacci 1988, p. 48].

Chapter 15
Maria Clara Eimmart (1676–1707)

> *The great men of science are supreme artists. (Martin H. Fischer)*

At the end of the seventeenth century, in an era when photography did not exist yet, and therefore it could not be used in support of astronomy, other kinds of skills were needed for this activity. For example, in Nuremberg, Germany one could have seen at night, on the roofs of the houses near the city walls, a woman observing at the eyepiece of a telescope. She is famous for her observations, but they are not of such kind that can fill in the catalogs with numbers. Rather, she needed to faithfully report what she was seeing with her drawings. Patiently, and unconcerned by the cold weather, night after night she composed these tables with comets, sunspots, eclipses, planets and especially with lunar mountains. Her name was Maria Clara Eimmart.

Born May 27, 1676, she was an astronomer and engraver, actually one of the first and most talented designer of astronomical tables, some of which are conserved at the Astronomical Observatory Bologna, Italy. Her mother was Mary Walther and her father Georg Chistoph Eimmart, the younger (1638–1705) a successful painter and engraver, as was his father Georg Chistoph Eimmart, the elder, a painter of landscapes, portraits and historical subjects.

© Springer International Publishing Switzerland 2016   97
G. Bernardi, *The Unforgotten Sisters,* Springer Praxis Books in Popular Astronomy,
DOI 10.1007/978-3-319-26127-0_15

Maria Clara Eimmart was born in a family of artists who also were active in the astronomical painting, as shown by this "Planisphaerium Coeleste", by her father

Maria Clara's father was born a Resenburg in 1638 and between 1655 and 1658 he studied mathematics at Jena, moving to Nuremberg in 1660 to reach his sister who, together with her husband, had founded an art studio. From 1699 to 1704 he was appointed director of the Academy of Arts in Nuremberg, the Malerakademie, but being also an amateur astronomer and since its main business was profitable, he was able to buy several astronomical instruments. He did not stop there and also built a private observatory which later became the city's Astronomical Observatory of Nuremberg and, the end of the century, the largest center of astronomical observation in Germany. Eventually he became a diligent observer publishing his results in different memories.

Maria Clara grew up immersed in this environment of art and science, thanks to her father she studied French, Latin and Mathematics, and learned her trade by specializing in botanical and astronomical illustrations. Also in this last area she learned her astronomical skills directly from her father, eventually becoming her assistant at their observatory, which soon became the meeting point for several astronomers. Among them she knew her future husband, Johann Heinrich Müller (1671–1731) professor of astronomy at Altorf. He became a student of her father and taught physics at the high school in Nuremberg.

They got married in January 20, 1706, and after the death of her father the Astronomical Observatory was purchased by the City of Nuremberg and Johann appointed as the new director. On May 12, 1706 Maria Clara observed the total eclipse from the Observatory, but the drawings made for that occasion went lost. The following year she ended her short life by dying in childbirth. The child did not survive.

15.1 Works

As it was common in other scientific fields, the astronomical treaties of the time were accompanied by appendices with illustrations, for example the work published in 1701 *Ichonographia nova contemplationum de sole in desolatis antiquorum philosophorum ruderibus concepta* (Ichonographia new contemplation of the sun) and authored by George Christoph Eimmart includes some illustrations by Maria Clara, although some argue that the entire work is of Maria Clara and went published under her father's name, however there is no evidence in support of this claim. The scientific purpose of these drawings is evident not only because of their accuracy, but also because the day of observation is recorded.

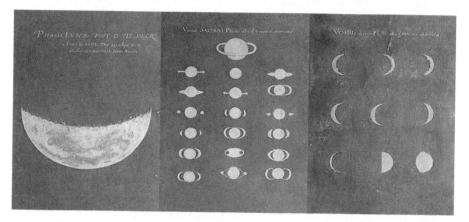

Astronomical illustrations by Maria Clara Eimmart

Some of the finest of the detailed illustrations of Maria Clara are several tables depicting different comets and planets like Mercury, Venus, Mars, Jupiter and Saturn, with 350 showing the phases of the Moon, which were made as the basis for a new lunar map. They were the result of careful telescope observations made between 1693 and 1698, an instrument that was becoming more and more widespread, and came soon to be an essential tool for a scientific community that could not use photographical plates yet. These boards were to illustrate her father's work *Micrographia stellarum phases lunae ultra 300* (Micrographia phases of the moon and the stars beyond 300) and are characterized by the blue sheets on which they are drawn with pastels.

15.2 Curious Facts

As mentioned above, the drawings that Maria Clara Eimmart realized during her short life did not concern exclusively astronomical subjects, but they also included flowers, birds, and ancient statues. Unfortunately all of them have been lost, except for one depicting a vestal that is kept at the National Museum of Germanicus, in Nuremberg.

Maria Clara's husband benefited from the marriage as well, because the Astronomical Observatory was part of her heritage, passing from the daughter to the husband, who became its director in 1705, the same year when the father died.

Ten of the works of Maria Clara can today be admired in Bologna, at the Observatory's Museum of the University. They are part of the twelve that arrived here as a gift to earl Marsili, a scientific collaborator of her father and citizen of this city, who we will also meet in a later chapter as a patron of two other female astronomers and who later donated them all to the observatory. The other two tables must have been missed some time after they became part of the belongings of the observatory, since the original statement of the donation of Marsili to the observatory reads: *"Tabulae XII. Chartacee ceruleo colore inductae, quibus caelestium corporum quoramdam Phases a Maria Clara Eimmart depictae sunt"* (Twelve tables. Blue colored sheets, which have been depicted by Maria Clara Eimmart with some phases of celestial bodies) thus testifying the presence of all of them at that time.

Here is the complete list of such drawings as reported by the observatory site (http://www.bo.astro.it/dip/Museum/english/car_67.html):

- several examples of the appearance of comets [Inv. MdS-124a];
- drawings of a paraselene and a parhelion (haloes around the Moon and Sun caused by reflections and refractions in the atmosphere) [Inv. MdS-124b];
- full moon [Inv. MdS-124c];
- lunar phase observed on 23 April 1693 [Inv. MdS-124d];
- lunar phase observed on 29 August 1697 [Inv. MdS-124e];
- phases of Mercury, according to the observations of Johannes Hevelius, of 1694, 1695 and 1696 [Inv. MdS-124f];
- phases of Venus [Inv. MdS-124g];
- aspect of Mars, according to the observations of various astronomers [Inv. MdS-124h];
- aspect of Jupiter, according to the observations of various astronomers [Inv. MdS-124i];
- aspect of Saturn, with view of the rings as they appeared in observations of the time [Inv. MdS-124l].

Three more tables add to the above ones. They are three studies of Lunar phases on brown paper, smaller than the previous, illustrating:

- crescent Moon observed at Nuremberg on 11 April 1681 [Inv. MdS-83a];
- waning Moon, observed at Nuremberg on 18 September 1695 [Inv. MdS-83b];
- crescent Moon observed at Nuremberg on 9 July 1695 [Inv. MdS-83c].

Chapter 16
Christine (1696–1782) and Margaretha (1703–1744) Kirch

The Kirchin

Christine and Margaretha are two of the daughters of the German astronomers Gottfried Kirch and Maria Margarethe Winkelmann-Kirch. Christine was born in Guben, Germany, in 1696. She was the eldest of the sisters and had an older brother named Christfried, born on 24 December 1694. Margaretha instead was born in 1703, and lost her father at the age of seven. All the sons and daughters of the Kirchs, since the age of 10, were instructed in what was the family business: astronomy.

16.1 Curios Facts

Christine, since a very young age and for most of her life, worked in the shadows for the father, the mother and then for her older brother and other assistants. It is known that since her childhood she helped her family mainly by taking the time or the time intervals with a pendulum, and later she was introduced in the realization of the calendars. In this task she started by helping her mother Maria Margarethe, and then her older brother Christfried. The latter, unlike his parents and his sisters, received an "institutional" education, in addition to that given within the family, in fact, up to 1712 he attended the prestigious Joachimsthalsches Gymnasium of Berlin, continuing his studies for 2 years in Nuremberg, then in Leipzig and later in Konigsberg. In 1710, after the death of Gottfried Kirch, the mother was issued some unsuccessful petitions to be officially recognized as a professional astronomer by the Academy of Sciences. After these vicissitudes Maria Margarethe Kirch and her daughters continued the same work as astronomers at other observatories. In 1715 Christfried rejoined his mother on her transfer to Danzig and worked here for 18 months at the famed Observatory of the deceased Johannes Hevelius, but shortly before, since the age of 20 he had already begun to publish annual planetary ephemeris. Maria Margarethe and her son received a job offer by Tsar Peter the Great after having shown him sunspots and other celestial phenomena from the

© Springer International Publishing Switzerland 2016
G. Bernardi, *The Unforgotten Sisters*, Springer Praxis Books in Popular Astronomy,
DOI 10.1007/978-3-319-26127-0_16

Observatory of Danzig, but the offer was declined, since the career of Christfried in Berlin was imminent. There in fact had recently died, in 1716, the astronomer Johann Heinrich Hoffmann, who had taken over Gottfried Kirch, after his death, in the directorship of the Observatory of Berlin, and the Academy of Sciences in Berlin offered Christfried a permanent position in astronomy, admitting him as a member of the Academy in October of the same year.

16.2 Works

At the Observatory of Berlin the mother and the sisters Christine and Margaretha worked with him as "assistants," but neither of his sisters nor their mother got official recognition for their profession. However, Christine finally received such recognition near the end of her career.

Carl Daniel Freydanck (1811–1887): "The New Observatory in Berlin", oil, 1838

At the Observatory the "Kirchin" continued to do astronomical observations and calculations for planetary ephemeris, and especially for the compilation of annual calendars issued by the Academy, which was the main activity required by the Academy. Here Christfried lived with his mother and sisters. He never married and in general the organization of the work at the Observatory remained basically the same as that of his father. More in particular, we know that, among the noticeable celestial phenomena, observations were made of the transit of Mercury in 1720, and the solar eclipse of 1733. Also, following a technique first suggested by Galileo a century before, the differences in longitude between Berlin, Paris and St. Petersburg was determined using the eclipses of Jupiter's satellites. In 1728 Christfried was promoted from the position of "observer" to that of regular "astronomer," and in 1723 he was admitted as a foreign member to the French Academy of Sciences and, after his death, to the Royal Academy of London.

16.3   The First Paid Woman Astronomer

Christine, however, did not have a regular salary for her work, and only in 1740 did she start to receive occasional and small donations from the Academy of Sciences of Berlin. This was because, after the death of his brother Christfried, due to a heart attack that occurred on March 9, 1740, the Berlin-Brandenburg Academy of Sciences and Humanities became more dependent on the professional help of Christine for making calendars. Moreover, in the same period, between 1740 and 1742, the King of Prussia, Frederik the Great conquered the new populous province of Silesia, an historical region of Central Europe today belonging almost entirely to Poland.

Map of the Electorate of Saxony in 1648

This increased significantly the income of the Berlin Academy, which indeed depended on its monopoly on the calendars in Prussia, and Christine became responsible for preparation of the calendar of this region. Eventually, in 1776, she received a salary of 400 thalers (the German silver coin) by the Academy and continued her work until late in life, thus becoming the first female astronomer paid for her professional activity, a record that until recently was attributed to Caroline Herschel. When the German astronomer reached the age of 77 the Academy awarded her the honor of "Emeritus," and in a letter addressed to her the Academy explicitly expressed its gratitude for her work on the calendars. She then continued to receive her salary, but without a binding working duty. Another indirect contribution of Christine was the introduction of a new astronomer of Berlin to the implementation of the calendars: the famous Johann Bode. There was a family relationship between the two, since Bode in 1774 had married a grandniece of Christine, and after the death of his first wife in 1782, he married the following year another grandniece of Christine, who was actually the older sister of his first wife. On the other side, very little is known about the life and the professional activity of the other younger sister, Margaretha Kirch, at the Berlin Observatory. It is known that she made cometary observations, in particular of comet 1743 C1, discovered on February 10, 1743 also in Berlin by the astronomer Augustin Grischow. She died at 41, the following year, while Christine survived her for several more years, dying in Berlin on May 6, 1782.

Chapter 17
Gabrielle Émilie Le Tonnelier de Breteuil, marquise du Châtelet (1706–1749)

Judge me for my merits
My youngest daughter is a strange creature destined to
become the most home-loving woman. If it were not for the
low opinion I have of many bishops I would route her to a
religious life and letting her hidden in a convent. She is twice
the height of a girl of her age, with a prodigious force, like
that of a lumberjack, and she is clumsy beyond any
imagination. Her feet are huge, but they can be easily
forgot as soon as you notice the huge hands.

This is what Louis-Nicolas Le Tonnelier de Breteuil, Baron of Preuilly, and head of ceremonies at the French court wrote about her daughter Gabrielle Émilie. Her mother, Gabrielle-Anne de Froulay, who belonged to the ancient military aristocracy, gave her birth in Paris on December 17, 1706.

Gabrielle Émilie Le Tonnelier de Breteuil, marquise du Châtelet at her desk

© Springer International Publishing Switzerland 2016
G. Bernardi, *The Unforgotten Sisters*, Springer Praxis Books in Popular Astronomy,
DOI 10.1007/978-3-319-26127-0_17

Gabrielle Émilie had six brothers and a half-sister, Michelle, born out of wedlock, whose mother, Anne Bellinzani was an intelligent woman and that became interested in astronomy and married to an important Parisian officer. Her father, in addition to the prominent position at the court of King Louis XIV, held a weekly salon on Thursdays, to which well-respected writers and scientists were invited.

Given the physical description made by the father, it is not surprising that her mother would have rather seen her daughter in a convent, as it was customary at the time, but the father recognized his daughter's intelligence and hired tutors who trained her in many languages like Latin, Greek, Italian, Spanish, German and English as well as in Mathematics, which became her major interest, even if she also liked dance, opera singing, fencing and riding.

Unlike her peers from the same social class who were educated in convents in expectation of a good marriage, Gabrielle Émilie spent her childhood and youth in the family home surrounded by governesses, the best tutors, and guests. One of them was the secretary of the French Académie des Sciences Bernard de Fontenelle, with whom Gabriella Émilie's father arranged a visit and a talk on astronomy when she was 10 still years old.

On June 12, 1725, at the age of 19 the young Gabrielle married Florent-Claude, Marquis du Châtelet and Earl of Lomont, who was 15 years her older. This union, like many at the time, was organized, and as a wedding gift her husband was appointed by Gabrielle Émilie's father Governor of Semur-en-Auxoisin Burgundy, where the couple moved.

They had three children: Françoise Gabrielle Pauline, born 30 June 1726, Louis Marie Florent, born 20 November 1727, and Victor-Esprit born 11 April 1733 who died in late summer 1734.

The Marquis du Châtelet was often absent because of his military duties, and after the three sons, Gabrielle Émilie, went back in Paris, giving herself to a light upscale living and resuming her mathematical studies. On algebra and calculus she was initially assisted by Pierre-Louis Moreau de Maupertuis, a member of the Academy of Sciences, who brought to her knowledge the work of Newton, and then by Alexis Claude Clairaut, who got his notoriety for an equation and a theorem bearing his name. As it was customary in the high society of the time, during these years she became involved into extramarital affairs with nobles or scientists, always tolerated and never concealed, but the most important one was with the philosopher Voltaire.

In the frontispiece to Voltaire's interpretation of Isaac Newton's work, Elémens de la philosophie de Newton (1738), the philosopher sits translating the inspired work of Newton. Voltaire's manuscript is illuminated by seemingly divine light coming from Newton himself, reflected down to Voltaire by a muse, representing Voltaire's lover Émilie du Châtelet—who actually translated Newton and collaborated with Voltaire to make sense of Newton's work

The marquise du Châtelet had met him years before, at the salon of her parents. Now she was 27 and he, 12 years older, was returning as a famous personality from his exile in England, and after having wandered across all Europe. In 1734 they went to live in the Château de Cirey, in Champagne, owned by the Marquis.

Château de Cirey

It was a perfect location because of its proximity to the border with Lorraine which for the defiant philosopher, frequently in danger of being arrested, was a strategic asset for a possible immediate escape. He took care of the refurbishment and modernization of the castle, and the two lovers remained here for nearly a decade to work in tranquility and receiving the high society of the time. The place became one of the brightest centers of the French philosophical and literary life of the time. The different interests of the Marquise, in fact, spanned from literature to geometry, physics, mathematics and astronomy, and with Voltaire she deepened various subjects, discussing in English and producing various works. The life at the castle followed precise and almost monastic rules: the morning was dedicated to the study, at 11 was served a meal, followed by a half-hour conversation, then everybody retired and assembled again around 9 p.m. for dinner.

17.1 Works

As an intellectual rather than a scientist, Voltarie implicitly acknowledged her contributions to his 1738 *Elements of the Philosophy of Newton*, where the chapters on optics show strong similarities with the *Essai sur l'optique* by the Marquise who, in addition to this, further contributed to the campaign by laudatory reviews in the *Journal des savants*.

Because of their common passion for science, they set up a laboratory which they both used, engaging themselves in a scientific collaboration, that also had a distinctive and healthy competitive character. Their independent participation in the 1738 Paris Academy prize contest on the nature of fire can be seen in such a context. Gabrielle Émilie and Voltaire disagreed in their essays, and although neither of them won, both received honourable mention and were published.

Title Page from "Dissertation sur la nature et la propagation du feu" by the marquise Du Châtelet

She thus became the first woman to have a scientific paper published by this Academy.

Among her scientific works we can find the *Institution de physique* (Lessons in Physics) a physical book meant for her son, about the Physics theory of Leibniz published in 1740. This disappointed Voltaire and was among the causes of breakage of their relationship.

In 1745 Gabriella Émilie began what was to become the greatest work of her life, the translation and commentary of the *Philosophiae naturalis principia Mathematics* that Newton published in 1687, the book where his laws of motion and gravity are exposed. That marked the beginning of a long, unstinted work that did not stop even when, at the age of 42, she got pregnant with her latest lover. She had the feeling of not being able to complete the work, and therefore worked day and night to finish it before the birth. Actually, she died of puerperal fever 6 days after the delivery, on September 10, 1749 in Lunéville. The daughter, Stanislas-Adélaïde, died in 1751, less than 2 years later.

The final edition consisted of two volumes, the first one and the first half of the second is the translation from Latin into French of Newton's *Principia*. The remaining part is a kind of summary of the opera, in about a hundred pages written entirely by Gabrielle-Émilie and entitled "*Exposition abrégée du Système du monde et explication des principaux phénomènes astronomiques tirée des Principes de m. Newton.*" (Abridged exposition of the World System and explanation of the main astronomical phenomena extracted from the Principia by Mr. Newton.)

Ten years after her death, Clairaut published the book. This was the only French translation at the time, and contributed decisively to the diffusion of the Newtonian philosophy in France, where the influence of the Cartesian theory was still significant, and still remains a reference today.

The philosopher Voltaire remembered her in this way: "She was a great man whose only fault was being a woman. A woman who translated and explained Newton [. . .] in a word, a really great man."

Here the list of her main works:

- *Dissertation sur la nature et la propagation du feu* (1st edition, 1739; 2nd edition, 1744.
- *Institutions de physique* (Paris, 1st edition, 1740; 2nd edition, 1742.
- *Principes mathématiques de la philosophie naturelle par feue Madame la Marquise du Châtelet* (1st edition, 1756; 2nd edition, 1759.
- *Principes mathématiques de la philosophie naturelle par M. Newton* translated from the Latin by Mme la marquise du Châtelet, accompanied by Commentaries on world system by M. Newton (Paris, Desaint et Saillant, 1756. Final edition 1759.

And not scientific:

- *Examen de la Genèse*
- *Examen des Livres du Nouveau Testament*
- *Discours sur le bonheur*

17.2   As They Said of Her

Voltaire's admiration for Gabrielle Émilie was boundless and the already cited declaration on her greatness (*"a great man whose only fault was being a woman"*) is contained in a letter to his friend King Frederick II of Prussia.

Among her admirers we can count the German philosopher Immanuel Kant, who commented that a woman *"who conducts learned controversies on mechanics like the Marquise du Châtelet might as well have a beard."*

For her part, Gabrielle Émilie exhorted the world to *"judge me for my merits"*, and not her sex.

Chapter 18
Maria Gaetana Agnesi (1718–1799)

The witch of Agnesi

A regular face, proportionate noise and a piercing glance. These are the most remarkable things that can be noticed in the portrait of Maria Gaetana Agnesi. A young woman with pearl earrings and elegant clothes probably made of silk, as her father was a silk merchant.

Maria Gaetana Agnesi by Bianca Milesi Mojon 1836

She was born in a wealthy family in Milan on 16th of May 1718 and she is mentioned in the poem "Letter of Caroline Herschel" by Siv Cedering as a scientist and astronomer. She was the oldest of 21 brothers and the family' salon was often crowded with the intellectuals of the time.

© Springer International Publishing Switzerland 2016
G. Bernardi, *The Unforgotten Sisters*, Springer Praxis Books in Popular Astronomy,
DOI 10.1007/978-3-319-26127-0_18

It became very soon evident that Maria was a child prodigy, and her father effectively encouraged her progresses, supporting her studies with the best teachers of the time. She studied philosophy, mathematics and astronomy, but she was also so talented for languages to be dubbed with the name of *Seven-tongue orator*. Indeed, she spoke Italian, German, French, Latin, Greek, Spanish and Hebrew, and in these languages she used to entertain her father's guests. Mary began since a very young age to participate in the cultural gatherings organized in her family's home, engaging with their guests in philosophical and mathematical discussions.

In 1738 a collection of more than one hundred philosophical and scientific essays was published in Milan with the title *Propositiones philosophicae quas crebris disputationibus domi habitis coram clariss. viris explicabat extempore et ab obiectis vindicabat M. C. de A. mediolanensis*. This work, written in Latin, reveals the broad range of scientific interests of Maria, who touched here several subjects, like logic, pneumatology, fluid and celestial mechanics. From meteors to fossil, from animals to metals, all the essays were exposed with an encyclopedic approach. Even before the publication of this book, which was inspired by the discussions heard during the gatherings of her father's salon, she had shown a distinct talent for mathematics, pursuing the study of the *Traitè analitique des secrions coniques* of the Marquis de l'Hopital, and about which she wrote an extensive commentary, never published but collected in one of the 25 volumes of unpublished works at the Ambrosiana Library of Milan. Later she continued to deepen her studies on mathematical analysis, and when her mother died she had the excuse to retire from public life taking over the management her large family.

18.1   Works

When in her twenties Maria started to compose a very important mathematical work on differential integral calculus, seemingly with the intent to teach this subject to his brothers. In 1748 it was published in Italian under the title of *Instituzioni analitiche ad uso della gioventù italiana* (Analytical Institutions for the Use of Italian Youth) and dedicated to the Empress Maria Theresa of Austria, since Milan at that time was part of the Austrian Empire, ruled by the Habsburg dinasty.

Frontispiece of the "Instituzioni analitiche" by Maria Gaetana Agnesi

As it can be read in the preface, Maria deliberately chose to use the Italian language instead of Latin, an indication from the very beginning that significantly highlights a distinct open-minded attitude, for those times, of the author toward the problems of education. This work was a great success and one of the first and most complete works on calculus as well. With respect to other treatises, in fact, this one was considered the work of various mathematicians in a systematic way, which was exposed along with the author's interpretations. It was organized in two volumes. The first one contained only one section (or "book", in the author's words) dealing with the analysis of finite quantities. It also includes the treatment of elementary problems of maxima, minima, tangents, and inflection points solved with algebraical methods. The second volume contains three sections/books. The first one deals with differential calculus, intended as the analysis of infinitely small quantities, and explains the methods of solving the same problems of maxima, minima, tangents and inflection points with this more general technique. The second section is about integral calculus, and gives a complete exposition of its now well-known applications to the problems of finding length of curves, areas of surfaces and volumes enclosed in solid figures. The last section deals with the so-called inverse method of tangents, that is of solution of first and second order differential equations. Being a model of clarity, it was translated into French and English and widely used as a textbook. Thanks to this work she was elected member of the Academy of Sciences of Bologna, which also offered her the chair of mathematics at the University of Bologna under the patronage of Pope Benedict XIV, but she never took the job. She was a very religious woman, and after the death of her father she entirely devoted herself to charity activities, transforming part of her home into a hospital. She also used valuable gifts of hers to support this activity, like a diamond ring given by the Empress Maria Theresa of Austria as recognition for her work that was sold to open another venue for her hospice. Later in her life she was given an appointment as director of the newly founded Pio Albergo Trivulzio in Milan, where she also took care of poor sick people, and especially of women, where she died penniless on January 9th of 1799.

18.2 Curious Facts

Because of an erroneous interpretation, the analytic function named *versiera of Agnesi*, was translated as *Witch of Agnesi* in the English translation of the *Analytical Institutions*.

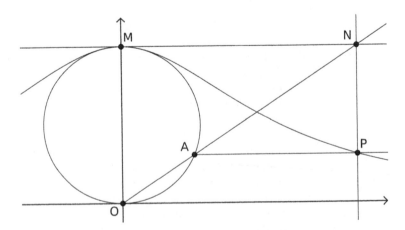

The mathematical curve named after Maria Gaetana Agnesi (the "Witch of Agnesi")

We can see in the figure a plot of this curve taken from the original work of Maria. Originally mentioned by Fermat almost 80 years before, the study of this geometrical object had been deepened by the Italian mathematician Guido Grandi in the early eighteenth century, who called it *versiera*. The name came from the geometrical procedure used in these first works to draw a curve. This procedure required one to move a point along the diameter of a circle, and this point identified a *versed sine* ("seno verso" in Italian) in this circle. All these words come from the Latin verb *vertere*, meaning "to turn," a concept that in the Italian of those times is well rendered by the name *versiera*. However, this term was also an abbreviation for *avversiera*, another Italian word meaning "the adversary [of God]," that is a "witch," (since avversiera is a feminine name) thus explaining how such a misinterpretation could have happened.

Another curious fact about this curve stands on its mathematical expression. In the *Analytical Institutions* its equation is $y = a \ \mathrm{sqrt}(a \ x - x^2)/x$, which is equivalent to $xy^2 = a^2(a-x)$,, while today's expression reads $yx^2 = a^2(a-y)$, switching the roles of x and y. This can be probably ascribed to a sort of "cultural change of habit" that happened in mathematics since the times of Guido Grandi and Maria Agnesi. As already mentioned above, they used a geometrical procedure to build this curve which made it more natural for them to consider the x-axis to be the vertical axis and the y-axis to be the horizontal one. Today, on the contrary, we are used to the concept of function which, in this case, requires that we use the opposite convention, that is x horizontal and y vertical.

There are quite a number of details suggesting that Maria's father was the inspiration, or even the main driver of her interest in mathematics. After the success of her book, Maria was made a member of the Academy of Sciences of Bologna, and in 1750 the university sent her a diploma, adding her name to the faculty and offering her the chair of mathematics. Despite these recognitions from the academic world, she apparently made little attempt to pursue a career in this field, or at least to furthering her activity in mathematics. In those years she had started to devote herself to the charity work which later will become her main interest, and although the appointment was not refused, she never took the chair. Moreover, when her father died in 1752, Maria gave up any further work in mathematics and increased her work in charity, and in 1762, at a request of the University of Turin asking for her opinion about the young Lagrange's recent articles on the calculus of variations, she replied that she was no longer concerned with such interests. Finally, one can read in the preface of the *Analytical Institutions* that she didn't want to take the burden of translating her work in Latin. This language was the official Academic language of the time, and since Maria was fluent in it, this decision cannot be ascribed to the difficulty of the task, and might be interpreted as a first, early indication that the young woman considered mathematics just as a temporary hobby of hers, rather than a professional interest.

18.3   As They Said of Her

Several witnesses were among her contemporaries, a sign that her reputation was enormous. In 1739 she was cited by the humanist Charles de Brosses as a living phenomenon, comparing her (not without a touch of irony) to Pico de la Mirandola for her extraordinary knowledge of languages and her somewhat encyclopedic versatility. In his *Lettres sur l'Italie,* describing Maria Gaetana as "a girl of about 20 years of age, neither ugly nor pretty, with a very simple and sweet manner". He went on to state that she expressed a particular interest in the work of Isaac Newton, but that she did not enjoy public discussion of this nature, "where for every one that was amused, 20 were bored to death".

Carlo Goldoni received from Maria Gaetana a copy of the *Istitutioni* and in return the famous playwright inserted a scene in the second Act of his comedy "Il medico olandese" (The Dutch doctor) staged for the first time in Milan in 1756, where he emphasized the fortune of this mathematical work in the Netherlands, where it was read and appreciated.

According to Dirk Jan Struik, Maria Gaetana Agnesi is "the first important woman mathematician since Hypatia".

There is a crater on Venus and an asteroid 16765 Agnesi (1996).

Chapter 19
Nicole-Reine Étable de la Brière Lepaute (1723–1788)

La savante calculatrice

Nicole-Reine Étable de la Brière was born on January 5, 1723 in Luxembourg Palace, then the residence of the regent of the kingdom of France Philippe d'Orléans, a member of the French royal family and nephew of Louis XIV. Her parents lived there because his father had worked for this family since a long time, first as a valet of the famous duchesse de Berry, and then for her sister Louise-Élisabeth d'Orléans. Already in her childhood Nicole showed her lively intelligence and a tireless zeal for the study, passing her nights reading books that she literally "devoured." On August 27, 1748 she married Jean André Lepaute, the royal clockmaker.

Portrait of Nicole-Reine Lepaute

© Springer International Publishing Switzerland 2016
G. Bernardi, *The Unforgotten Sisters*, Springer Praxis Books in Popular Astronomy,
DOI 10.1007/978-3-319-26127-0_19

The young couple lived in the Luxembourg Palace, and Nicole's husband became famous all over Europe for the considerable technical achievements in his field. For instance, he is credited with the realization of the first ever made horizontal clock, and of a clock with a single wheel. Nicole soon began to collaborate with her husband, her early research concerning the swing of pendulums of different lengths, but this activity eventually lead to deepen her interest for Astronomy. The genesis of this interest has probably to be traced back to the influence of a very young Joseph Lalande, at the time pupil of the astronomer Joseph-Nicolas Delisle, whom some sources indicate as responsible for the local astronomical observatory at the palace. It would have been quite natural for a young astronomer to take advantage of a fine craftsman who could provide the accurate clocks needed for his observations, and eventually many astronomical observatories in Europe were provided with Lepaute's pendulum clocks. In further support, it is attested that in 1753 Lalande, then a new member of the Academy of Sciences of Paris, examined and approved an innovative clock sent to this institution by Nicole's husband for evaluation. The work in this field lead to a book entitled *Traité d'horlogerie* written by Jean André Lepaute in 1755 and to which, as reported by Lalande, Nicole actively collaborated. Her first important work in Astronomy can be dated back to 1757, when she started collaborating with Lalande and Alexis Clairaut, who we already know for his collaboration with the Marquise du Châtelet. She became more and more involved in Astronomy, also as author of numerous and appreciated personal works.

Later in her life, in 1768, a nephew of her husband named Lepaute d'Agelet came from Montmédy to become an astronomer. He was fifteen then and 5 years later, in 1773, he participated in the journey of the explorer Yves de Kerguelen in the southern hemisphere, which earned him a professorship of mathematics at the École Militaire of Paris, and later access to the Academy of Sciences in 1785. Nicole will instead fail in her purpose to attract another nephew toward astronomy since, although he was gifted in mathematics, his parents preferred to orient him towards a career as a notary.

Generous and unselfish, she passed the last years of her life nursing her sick husband, and died a few months before him, on December 6, 1788. In the same year, before her death, she had been admitted to the Académie des Sciences of Beziers.

19.1 Works

As reported in the following list of her main works, M.me Lepaute's principal activity was as an astronomical calculator:

- *Table des longueurs des pendules, in Traité d'horlogerie, 1755*
- *Observations, in Connaissance des temps, 1759–1777*
- *Carte du passage de l'ombre de la Lune au travers de l'Europe dans l'éclipse annulaire du Soleil qui doit arriver le 1er avril 1764,* Paris, 1762
- *Table des angles parallactiques, in Connaissance des temps,* Paris, 1763
- *Table du Soleil, de la Lune et des autres planets, in Ephémérides du movement celeste,* de Lalande, 1774
- *Mémoires d'astronomie, in Mercure.*

Before recoiling from the diminutive sense which would be natural for our days, we must remember that at that epoch computers did not exist and their mechanical ancestors were rare and of a quite limited use. For this reason astronomical calculations were a common and essential activity of all scientists of the time, as we will see in short from Lalande's report of their collaboration. Actually, Nicole's work as a calculator has therefore to be considered at the same level as that of her colleague scientists, which required non-trivial physical and mathematical knowledge, rather than a mere "mechanical" application of some algorithms.

We can start from the first work on her husband's *Traité d'horlogerie,* about which Lalande, in his *"Bibliographie astronomique"* reported that *"Madame Lepaute is immediately entered into this working group, she was too smart not to have the curiosity; she noted, she calculated, she described the work of her husband. We have embarked on a new common Traité d'horlogerie which appeared in 1755, printed in 4th."* We can thus start understanding what was meant in our previous statement about the activity of a calculator.

xx

TABLE VI.

De la longueur que doit avoir un Pendule simple pour faire en une heure un nombre de vibrations quelconque, depuis 1 jusqu'à 18000.

Calculée par Madame LEPAUTE.

Nombres de vibrations par heure.	pieds.	pouces.	lignes.	Décimales, ou centièmes de lignes.	Nombres de vibrations par heure.	pieds.	pouces.	lignes.	Décimales, ou centièmes de lignes.
18000	0	1	5	62	15100	0	2	1	04
17900	0	1	5	82	15000	0	2	1	38
17800	0	1	6	02	14900	0	2	1	72
17700	0	1	6	22	14800	0	2	2	07
17600	0	1	6	43	14700	0	2	2	42
17500	0	1	6	64	14600	0	2	2	78
17400	0	1	6	80	14500	0	2	3	16
17300	0	1	7	08	14400	0	2	3	53
17200	0	1	7	30	14300	0	2	3	92
17100	0	1	7	52	14200	0	2	4	32
17000	0	1	7	70	14100	0	2	4	72
16900	0	1	7	99	14000	0	2	5	13
16800	0	1	8	24	13900	0	2	5	55
16700	0	1	8	47	13800	0	2	5	98
16600	0	1	8	72	13700	0	2	6	42
16500	0	1	8	97	13600	0	2	6	87
16400	0	1	9	23	13500	0	2	7	33
16300	0	1	9	49	13400	0	2	7	80
16200	0	1	9	75	13300	0	2	8	28
16100	0	1	10	02	13200	0	2	8	77
16000	0	1	10	30	13100	0	2	9	27
15900	0	1	10	59	13000	0	2	9	79
15800	0	1	10	87	12900	0	2	10	31
15700	0	1	11	16	12800	0	2	10	85
15600	0	1	11	46	12700	0	2	11	40
15500	0	1	11	76	12600	0	2	11	96
15400	0	2	0	07	12500	0	3	0	54
15300	0	2	0	39	12400	0	3	1	13
15200	0	2	0	71	12300	0	3	1	74

TABLE VI.

Pendulum's table from the Traité d'horlogerie (1767) by Jean-André Lepaute calculated by his wife Nicole-Reine

This remarkable work includes a description of the equations of the pendulum. Calculation of the tables made by Lepaute's wife required the understanding of these equations and of the ideas adopted to describe and compute the reported differences between the true and the mean time. In the same book Lalande adds also that *"Madame Lepaute computed for this book a table of numbers of oscillations for pendulums of different lengths, or the lengths for each given number of vibrations, from that of 18 lignes, that does 18000 vibrations per hour, up to that of 3000 leagues."* This quite obscure sentence refers to Table VI of the *Traité*, which reports the length of a pendulum (in feet, inches, "lignes" and hundredths of lignes) corresponding to a specific number of oscillations per hour. The first one is that required to have 18,000 oscillations per hour, then it goes at intervals of 100 down to 1 oscillation per hour, which would require a length of about 3000 leagues, or 12,000 km.

The passion of Madame Lepaute for mathematics and astronomy led her to participate in an even more ambitious scientific adventure. In June 1757 the astronomer Lalande resolved to accurately determine the date of the return of Halley's comet, foreseen for the following year, or an accurate confirmation of Edmond Halley's forecasts made at the beginning of the century. To this aim it was necessary to calculate the effects of the gravitational pull of Jupiter and Saturn on the path of the comet. The French astronomer involved in this task also the above mentioned mathematician Alexis Claude Clairaut, who had found a solution for the calculation of the three-body problem. It was a huge job that required calculation of the distances and forces of attraction exerted on the comet by each planet by using Newton's laws, and at intervals of one grade, and for a time period of 100 and 50 years. Moreover, they obviously needed to complete this work before the returning of the comet. So, for more than 6 months the three scientists calculated relentlessly, to the point of getting sick. So colossal was the amount of work needed that still today, with computers at our disposal, it seems incredible that this project could have been carried out by hand. As once again reported by Lalande: *"For six months we made calculations from dawn to dusk, sometimes even during the meals [...] The help given by M.me Lepaute was such that without her I would not have been able to complete such a colossal enterprise, [...]"* This excerpt shows even better than before how the computations were performed by all the scientists who didn't regard them as a mere mechanical work and, on the contrary used them to collect praise among their peers. It is then noticeable that Nicole won in this way the reputation of being one of the best astronomical calculators. On November 14, 1758, together with other astronomers, they reported the Academy of Sciences the dates of the passage of the comet, predicting a periodicity of about 75 years. The three astronomers announced that the comet would have reached the perihelion, i.e. the point of the comet orbit closest to the Sun, not in 1758 as foreseen by Halley, but in mid-April 1759, with an error margin of a month. Actually, the comet crossed the perihelion on March 12, 1759. This was a breakthrough because it was the first time that scientists were able to predict the perihelion passage of a perturbed comet: just in time to be spotted as was observed on December 25th.

Lalande will pay tribute to Madame Lepaute for her important contribution in his book *Théorie des comètes*, while Clairaut, that was conscious of her merits, called her the "*savante calculatrice*," the learned calculator. However, he was not explicit in his praise, a fact that Lalande ascribes to "*[. . .] the compliance of a woman jealous of the merit of M.me Lepaute. and who had pretensions without any kind of knowledge. She drove to commit an injustice a wise but weak scholar, whom she had subdued.*" So in his *Théorie du mouvement des comètes* Clairaut first recognized the work of his female colleague, but he then excised it from the printed version and later denied her contribution, claiming for himself the exclusive merit of the calculations.

Nonetheless, Madame Lepaute continued her mathematical and astronomical activity. In 1761, she published the calculations of all the observations made during the transit of Venus across the Sun. In 1762 she computed the relevant data of a new comet and calculated in advance the duration and the size of an annular Solar eclipse that would have occurred in 1764 in Europe. This work was published in a book entitled *Explication de la carte qui représente le passage de l'ombre de la lune au travers de l'Europe dans l'eclipse du soleil centrale et annulaire di 1 Avril 1764 presenté au Roi, le 12 août 1762, par Mme Lepaute*. (Explanation of the map representing the passage of the shadow of the Moon across Europe during the central and annular Solar eclipse of April 1, 1764 presented to the King on August 12, 1762, by M.me Lepaute.) She had determined here the duration and the percentage of the eclipse visible in different European countries and showed in two maps, one for the whole Europe and the other for Paris, the progress of the eclipse at intervals of 15 min and its different phases. In the *Journal des savans* of April 1769 Lalande wrote: "*It can be seen on the map of Madame Le Paute the trace of the shadow that formed on the Earth a small ellipse which ran a space of twelve leagues per minute; this speed is twice that of a cannon ball, which is about two hundred arms per second, or five leagues per minute.*" Also, thanks to the data collected from several eclipses, Madame Lepaute prepared a table of parallax angles that appeared in the *Connaissance des temps* of 1763 and in a work entitled *Exposition du calcul astronomique*.

The *Connaissance des temps* was an annual publication of the Academy of Sciences for astronomers and navigators issuing the positions of the different celestial bodies for each day of the year. When Lalande became in charge of this publication, it was still Madame Lepaute who dealt with this huge computational work. This continued until 1774, when she took over the responsibility of the *Ephé mérides* of the Academy, for which she had to compute the positions of the planets, the Sun and the Moon. She went on for 10 years, and the eighth volume, published in 1783, contained the data until 1792.

19.2   Curious Facts

When she was a child one of her sisters said to her "*I am the whitest*" and Nicole replied: "*And I, the wittiest!*"

The astronomer Lalande, grateful for so much work and perhaps also susceptible to the charm of Madame Lepaute, dedicated to her some verses where she is called *Sinus des Graces*—which in French can be interpreted as "Sine of graces" but also "Bosom of graces"—and *Tangent de nos coeurs*—also a word pun between a mathematical "tangent" and a "touching" of "the hearts." She herself was certainly not indifferent to them.

Lalande had a peculiar habit: he ate delicious spiders and caterpillars and bragged about it. Nicole instead was afraid of these insects, but for him she got used to see them, touch them and, finally, to swallow them.

Unfortunately, the years of arduous astronomical calculations made her almost blind.

Modern astronomers have paid tribute to the talents of Madame Lepaute by naming a crater of the Moon after her.

The Luxembourg Palace, birthplace Nicole Lepaute was built by Maria de' Medici and used as the ancient seat of the kings of France. Today it is home to the French Senate since 1958. The large gardens were opened to the public in 1778.

19.3   As They Said of Her

Joseph-Jérôme Lalande, after her death, wrote about Madame Lepaute in his work *Histoire de l'astronomie abrégée*: "*This interesting woman is often in my thoughts, always dear to my heart; the moments that I spent close to her and her family are the ones I most love remembering, and whose memory, mixed with bitterness and pain, spread some sweetness on the last years of my life, as his friendship was the charm of my youth. Her portrait that I always have under my eyes, is my comfort, when I think that a philosopher should not complain about the laws of the imperious necessity, and of the losses that are a necessary consequence of the order of nature.*"

Part IV
Timelines from Louise Elisabeth Félicité Pourra de la Madeleine Du Piérry to Mary Fairfax-Somerville

1800 Alessandro Volta invents the battery

1801 Giuseppe Piazzi discovers the first asteroid

1814 Fraunhofer observes the Solar spectrum

1815 battle of Waterloo

1825 Laplace publishes *Celestial Mechanics*

1823 Babbage builds the analytical engine, precursor of computers

1838 Bessel measures the first parallax

1844 Bessel announces that Sirius is part of a double star system

1846 Le Verrier discovers Neptune

1850 first photograph of a star

1859 Darwin publishes the book enunciating the theory of evolution of species by natural selection

1861 the American Civil War

1869 the chemist Mendeleev presents the first version of the periodic table

1874 Heinrich Schliemann announces the discovery of the city of Troy

1900 with his explanation of the blackbody law the physicist Planck kicks off Quantum Mechanics

Chapter 20
Louise Elisabeth Félicité Pourra de la Madeleine Du Piérry (1746–?)

> *[...] elle représente un modèle pour toutes les femmes à cause de ses hautes qualités intellectuelles. (Joseph-Jérôme Lalande, Astronomie des dames)*

Louise Elisabeth Félicité Pourra de la Madeleine was born on August 1, 1746 in La Ferte-Bernard, a small city of the ancient French province of Maine. She became interested in science at a very young age. When she married Monsieur Du Pierry at age 20, she decided to continue her pursuit of a scientific career.

Joseph-Jérôme Lalande, director of the Paris Observatory, introduced the young lady to astronomy and asked her to collect historical data on the movements of the moon and of the eclipses that occurred in the last century. In this way a steady collaboration was started which brought her to produce a certain number of important works, mainly devoted to the production of astronomical tables.

© Springer International Publishing Switzerland 2016
G. Bernardi, *The Unforgotten Sisters*, Springer Praxis Books in Popular Astronomy,
DOI 10.1007/978-3-319-26127-0_20

The Paris Observatory at the beginning of the eighteenth century. It is interesting to notice the use of long tubed telescopes and even longer tubeless Aerial telescopes which were common instruments at the times

20.1 Works

Louise Elisabeth had a quite dynamic scientific activity, in consideration of the difficulties caused by her sex and the historical period. A list of her works includes:

- *Tables de l'effet des réfractions, en ascension droite et en déclinaison, pour la latitude de Paris*, Paris, 1791.
- *Tables de la durée du jour et de la nuit*, Paris, 1792.
- *Calculs d'éclipses pour mieux trouver le movement de la Lune*.
- *Table alphabétique et analytique des matières* continues dans le cinq tomes du *Système des connaissances chimiques* de Fourcroy, Paris, Beaudouin, an X.

The first one is about the light refraction caused by the atmosphere. When light passes from one mean to another one it deviates from its original direction of propagation. In astronomy, and especially in position astronomy, this effect has to be taken into account because it causes a displacement of the observed objects which depend on the thickness and the characteristics of the air crossed by its light. In other words, the effect is specific of the observing place and of the observing direction. The work of M.me Du Pierry therefore consisted in the estimation of the refraction effect, which had to be included in the calculations of the astronomers, which was published in a series of tables providing its amount as function of the right ascension and declination at the latitude of Paris.

The second book was another result of her long-standing collaboration with Lalande, where the duration of days and nights to be used for astronomical and civil purposes are reported.

The computation of the orbit of the Moon was an important scientific problem of that time, since it was used to investigate several connected phenomena. For example, it was needed to have precise models of the tides and for the eclipses prediction. The historical data of past eclipses, and their prediction and successive observation, which was the subject of the third work of Louise, therefore served to improve the knowledge of the Lunar orbit.

Her work was not confined to a research activity. Indeed Louise had been appointed Professor at the Sorbonne University, leading in 1789 an astronomy course for women who had enjoyed a great success. Recent research of Lémonon Waxin, instead, claim that her teaching activity was privately managed at home, rather than in public places, but in any case the French Revolution stopped that enterprise and Madame Du Pierry had to return to more discreet activities.

20.2 Curious Facts

Louise Elisabeth realized in 1799 (year X of the French revolution, as reported in the above bibliography) the *Table alphabétique et analytique des matières contenues dans les 10 tomes du système des connaissances chimiques* (Alphabetical and analytical tables of the materials contained in the 10 volumes of the System of chemical knowledge) for the chemist Antoine Fourcroy, a pupil of the famous Antoine-Laurent de Lavoisier, and State Chancellor in charge of public education. It is regarded as a remarkable work both in terms of importance, it consists of 170 pages printed in fourth and with small characters, as well as a masterpiece of precision and clarity.

Lalande enlisted the cooperation of several women for his work as an astronomer, but M.me du Pierry calculated most of the eclipses used by Lalande for the study of the motion of the Moon. His consideration for Louise was so high that he dedicated to her his book *Astronomie des dames*.

Madame Du Pierry ended her life forgotten at the beginning of the nineteenth century, but the precise year of her death is not known, although it seems ascertained that it was after the death of Lalande, in 1807.

She was a member of the Académie des Sciences de Béziers, for whose latitude she had calculated the duration of the day and the night.

20.3   As They Said of Her

In 1790, in his book *Astronomie des dames*, Lalande publicly expressed his praise for his female collaborator, stressing in particular her talent, her good taste and her courage: *"[. . .] elle représente un modèle pour toutes les femmes à cause de ses hautes qualités intellectuelles."* ([. . .] she represents a model for all women for her high intellectual qualities.)

In the same book he also reports that *"she has made several calculations on the eclipses to find the best lunar motion; she was the first woman who taught astronomy in Paris"* which unfortunately is not sufficient to clarify the claims about her professorship.

Chapter 21
Caroline Lucretia Herschel (1750–1848)

Eight comets for herself

Caroline Herschel could truly be regarded as a real-life Cinderella, in fact her fate had been sealed against her will. She had nothing to look forward to but being a kitchen maid at the service of her family, but not at the behest of the stepmother as in the tale, but for that of real mother herself, who thought it was useless that she be educated, as the father instead had wished. Carolina had not been favored by nature from the aesthetic point of view. She was not handsome, and suffered from a disease that delayed her body development and growth, so that for the rest of her life she remained of short height. Adding the fact that she had no dowry to offer meant a sure future as a maid. Luckily, a virtual prince charming came along when she moved away to be with her brother William, who was in England to pursue his musical career. This freed the young Caroline from a life of domestic duties and allowed her to also take up a career in music, in which she became a promising soprano. Later, the passion of William for astronomy took the upper hand on his musical interests and, thanks to the salary granted by the king in support of his research, he was able to totally devote himself to science.

© Springer International Publishing Switzerland 2016
G. Bernardi, *The Unforgotten Sisters*, Springer Praxis Books in Popular Astronomy,
DOI 10.1007/978-3-319-26127-0_21

Sir William Herschel and Caroline Herschel. Colour lithograph by A. Diethe, ca. 1896. Creative
Commons Attribution 2.0 Generic

His sister was dragged into his new adventure and, eclectic as she was, she not
only adapted to it, but also pushed herself into it with passion and commitment till
the end of her life.

21.1   From Cinderella to Soprano

Caroline was born on March 16, 1750 in Hanover, then one of the German states
of the Holy Roman Empire, now in Germany. The eighth child out of 10 of Isaac
Herschel and Anna Ilse Moritzen, she was given the second name of Lucretia, while
within her family she was called Lina. Actually Carolina had only four brothers and
one sister, because four died young; those who survived were Sophia Elisabeth, born
on April 12, 1733, Heinrich Anton Jakob, born on November 20, 1734, Friedrich
Wilhelm better known as William, born on November 15, 1738, Johann Alexander,
born on November 13, 1745 and Johann Dietrich, born on September 13, 1755.

Carolina's father was a gardener who became a military musician, namely an oboist in the band of the Hanover military band. Despite his poor education, he did his best to educate his four sons and two daughters, also bringing them closer to his interests such as music, philosophy and astronomy. The mother did not support her husband's attempts and, although she reluctantly accepted that her four children would all have a perfunctory education, she was strongly opposed to that for the daughters, which she said would have to deal only with housework. As Caroline herself wrote in her memoirs: *"My father wished to give me something like a polished education, but my mother was particularly determined that it should be a rough, but at the same time a useful one; and nothing farther she thought was necessary but to send me two or three months to a sempstress to be taught to make household linen. Having added this accomplishment to my former ingenuities, I never afterwards could find leisure for thinking of anything but to contrive and make for the family in all imaginable forms whatever was wanting, and thus I learned to make bags and sword-knots long before I knew how to make caps and furbelows.... My mother would not consent to my being taught French, and my brother Dietrich was even denied a dancing-master, because she would not permit my learning along with him, though the entrance had been paid for us both; so all my father could do for me was to indulge me (and please himself) sometimes with a short lesson on the violin, when my mother was either in good humour or out of the way. Though I have often felt myself exceedingly at a loss for the want of those few accomplishments of which I was thus, by an erroneous though well-meant opinion of my mother, deprived, I could not help thinking but that she had cause for wishing me not to know more than was necessary for being useful in the family; for it was her certain belief that my brother William would have returned to his country, and my eldest brother not have looked so high, if they had had a little less learning"*.

In addition to these prospects, it must recalled that the little Carolina was certainly disfavored by the events: at the age of 3 she contracted smallpox that left her left eye slightly disfigured, and at ten she suffered from typhus, remaining of short height for life because of its consequences. Indeed, she was 4′ 3″, or about 1 m and 40, and since she was neither handsome nor rich she certainly could not be regarded as a future bride. The father never hid these facts from her, and the mother was planning to use her as a maid for their large family. The inward pain that a teenager had to feel about what the future laid out for her can only be imagined.

The four brothers of Carolina started careers as musicians, while she showed an enthusiasm for science that her father tried to meet, despite her mother's efforts to ensure her a future as a housekeeper. In her youthful memories the future astronomer reported one about her father who one evening brought her: *"[...]... on a clear frosty night into the street, to make me acquainted with several of the most beautiful constellations, after we had been gazing at a comet which was then visible."*

Despite the astonishment Caroline felt that evening, she certainly could not imagine that 1 day she would have been engaged in this as a job, and even less her future discoveries. In 1756 the Seven Years' War broke out and 1 year later, when Carolina was 7, the state of Hanover fell under occupation by the French.

Isaac Herschel was engaged in war, and his son William, at 19, fled to England as a musician, becoming in 1766 an organist and choirmaster of the Octagon Chapel, in Bath, a fashionable resort of the wealthy in the eighteenth century. This was a town with many opportunities for a musician who could be often hired to give private recitals and music lessons. Seven years after his return, in 1767, the father died, and the eldest brother Jacob became the head of family. He was a brilliant musician, but also unsympathetic, demanding and liable to whip her if Caroline did not wait at table to his satisfaction. The young girl then realized that she had to take control of her own life to hope for a decent future, so she began to take classes in tailoring and studied to qualify as a governess. The turning point came in 1772, when William, called Fritz, feeling sympathy for Caroline and needing a housekeeper, offered her a place in his home in England, in spite of the protests of his mother. At 22, she joined her brother who began to give her singing lessons. She eventually won quite some success in her representations as first treble in the Messiah, Judas Macabaeus, etc., she sang at Bath and Bristol sometimes five nights per week, and she was invited to appear as a soloist in the Birmingham festival. Her answer would have determined a future direction of her life and for the history of astronomy. She declined the offer, having resolved to appear only where her brother conducted.

Caroline continued to collaborate with her brother on his official job as a musician, receiving for this activity English lessons, in addition to those of music. However William spent almost all of his free time studying astronomy and mathematics, devoting also many hours to improving his skills on the construction of telescopes. Word of the activities of the musician-astronomer was not unusual in English scientific circles, including some of William's friends as two professional observers, Thomas Hornsby of Oxford and Nevil Maskelyne, the Astronomer Royal. Caroline therefore started to follow him on this way, thus getting taught also in algebra, geometry and spherical trigonometry, useful for astronomical observations, up to the point of reading advanced texts such as the "*Fluxions*" of Maclaurin. She herself remembers that all the free time was taken feverishly up by his scientific interests: "*But every leisure moment was eagerly snatched at for resuming some work which was in progress, without taking time for changing dress, and many a lace ruffle was torn or bespattered by molten pitch, &c., besides the danger to which he continually exposed himself by the uncommon precipitancy which accompanied all his actions, of which we had a melancholy sample one Saturday evening, when both brothers returned from a concert between 11 and 12 o'clock, my eldest brother pleasing himself all the way home with being at liberty to spend the next day (except a few hours' attendance at chapel) at the tinning bench, but recollecting that the tools wanted sharpening, they ran with the lantern and tools to our landlord's grindstone in a public yard, where they did not wish to be seen on a Sunday morning.... But my brother William was soon brought back fainting by Alex with the loss of one of his finger-nails. This happened in the winter of 1775, at a house situated near Walcot turnpike, to which my brother had moved at midsummer, 1774. On a grass plot behind the house preparation was immediately made for erecting a twenty-foot telescope, for which, among seven and ten foot mirrors then in hand, one of twelve foot was preparing; this house offered more room for workshops, and a place on the roof for observing. [. . .]"*

The reflectors William wished for his research needed large mirrors which could not be found off-the-shelf. He then first decided to buy row disks and to grind and polish them by himself. Eventually the diameters of his telescopes became so large that even the row disks were beyond the capacities of the local foundries, so William started casting disks himself at home, and Caroline found herself spending long hours pounding horse-dung for the mould.

> For my time was so much taken up with copying music and practising, besides attendance on my brother when polishing, since by way of keeping him alive I was constantly obliged to feed him by putting the victuals by bits into his mouth. This was once the case when, in order to finish a seven foot mirror, he had not taken his hands from it for sixteen hours together. In general he was never unemployed at meals, but was always at those times contriving or making drawings of whatever came in his mind. Generally I was obliged to read to him whilst he was at the turning lathe, or polishing mirrors, Don Quixote, Arabian Nights' Entertainment, the novels of Sterne, Fielding, &c.; serving tea and supper without interrupting the work with which he was engaged,..... and sometimes lending a hand. [. . .]

Since 1781 astronomy became no longer a hobby for William because he became famous thanks to an important discovery. Initially mistaken for a comet it was instead a new planet: "Georgium sidus" (Georgian star) which afterwards was named Uranus. George III, King of England and Elector of Hanover (William was an Hanoverian resident in England), gave him a salary of 200 pounds a year for this, which was not too generous (as a musician he was earning 300), but enough to allow him to devote himself full time to astronomy. With a final performance in St. Margaret's Chapel, on Whit-Sunday, 1782, the musical career of William and Caroline came to a close. Caroline then abandoned her singing career with some regret, because music, in her opinion, was her true and only vocation; becoming involved in her brother's astronomical activity. The contemplation of herself in the guise of an assistant-astronomer moved her to cynical self-scorn. She helped him both in his observations and in the manufacturing of the instruments, and soon William gave her a telescope with which she began to observe the heavens on her own, conducting systematic surveys of the sky, and in particular seeking for those comets that fascinated her so much in her childhood.

21.2 Becoming a Professional Astronomer

In August of 1782 they moved to Datchet, near Windsor, in a dilapidated gazebo, because William needed a much larger place to realize his astronomical activity.

Caroline's new tole was mainly to help her brother with his astronomical projects: William in fact asked her to abandon her own observing activity because he had to be at the eyepiece without interruption, ready to call out a description of any nebula that came into view and she seated at the desk at an open window nearby, and nearly as cold as her brother, would write down his description, call it back to him for confirmation, and record the time of the observation and the altitude of the new 20 ft telescope.

During the day she had to work on the results obtained in the previous night, a task conducted with the extraordinary accuracy that is required to perform long calculations. In this way she had much less time for her own observations, but when her brother was not at home, she was able to recover her research program.

A curiosity is that Caroline never learned multiplication tables because she studied them so late in life, so she carried a table on a sheet of paper in her pocket when she worked, but yet a computational error has never, we believe, been imputed to her.

One of her interests was in particular, deep sky objects, and by the end of 1783 she had discovered fourteen objects, including galaxies and open clusters, which were included in William's catalog. It was, however, not an easy job, as testified by her diary, published by her nephew John Herschel: "[...] *I knew too little of the real heavens to be able to point out every object so as to find it again without losing too much time by consulting the Atlas. But all these troubles were removed when I knew my brother to be at no great distance making observations with his various instruments on double stars, planets, &c., and I could have his assistance immediately when I found a nebula, or cluster of stars, of which I intended to give a catalogue; but at the end of 1783 I had only marked fourteen, when my sweeping was interrupted by being employed to write down my brother's observations with the large twenty-foot. I had, however, the comfort to see that my brother was satisfied with my endeavours to assist him when he wanted another person, either to run to the clocks, write down a memorandum, fetch and carry instruments, or measure the ground with poles, &c., of which something of the kind every moment would occur. [...]*".

Two of the objects these words refer to are galaxies: NGC 253, discovered on September 23, 1783, and NGC 205, a member of the same local group of galaxies of Andromeda and the Milky Way, discovered on August 27, 1783.

NGC 253, also known as the Sculptor Galaxy as observed with the 1.5-metre Danish telescope at the ESO La Silla Observatory in Chile. Creative Commons Attribution 4.0 International license, ESO/IDA/Danish 1.5 m/ R. Gendler, U.G. Jørgensen, J. Skottfelt, K. Harpsøe

All the other objects are open clusters: NGC 7789, discovered in the fall of 1783 with an 80–100 mm diameter instrument, and three more (NGC 189, NGC 225, NGC 659) are in Cassiopeia. This star cluster, which contains about a thousand objects within a radius of 50 light-years at approximately 6000 light-years from here, is somewhat peculiar for an open cluster. This is because usually open clusters look like stellar gatherings only because they are formed by young stars born together in a nebula. Because of their young age they therefore have not had enough time to scatter away, but eventually they will do so since they are not gravitationally bounded. This usually happens in a few tens of million years, which therefore represents a typical limit for the age of an open cluster. That of NGC 7789, instead, is about 1.5 billion years. This can also be seen by their color, as its members are generally yellow and orange, while the million-years-aged stars which represent the typical population of an open cluster are usually more on the blue side. The list also includes two open clusters in the constellation of Cygnus (NGC 6819 and 6866), and one for each constellation in Andromeda, Cepheus, Canis Major, Hydra and Ophiuchus, that is respectively NGC 752, NGC 7380, NGC 2360, NGC 2548 and NGC 6633. The fourteenth object is also an open cluster in Ophiuchus, whose catalog number is IC 4665, as reported by Michael Hoskin in 2006. Its inclusion in Caroline Herschel's list happened after some debate because other observers shared the opportunity of cataloguing it, but none, with the possible exception of Caroline, did it to such a fully extent, until it received its final identification number in the second version of the Index Catalogue, in 1908.

21.3 Comet-Hunter

Reading from Caroline's memories, the idea of finding comets was not unknown to her: "*.... In my brother's absence from home, I was of course left solely to amuse myself with my own thoughts, which were anything but cheerful. I found I was to be trained for an assistant-astronomer, and by way of encouragement a telescope adapted for "sweeping," consisting of a tube with two glasses, such as are commonly used in a "finder," was given me. I was "to sweep for comets," and I see by my journal that I began August 22nd, 1782, to write down and describe all remarkable appearances I saw in my "sweeps," which were horizontal. But it was not till the last two months of the same year that I felt the least encouragement to spend the starlight nights on a grass-plot covered with dew or hoar frost, without a human being near enough to be within call. [...]*".

In April 1786 the Herschel's moved to Slough in a new home, which they called Observatory House and Caroline received here, from her brother, a small telescope with a focal length of about 70 cm capable of 30 magnifications. On the night of August 1 the 36 years-old astronomer discovered her first comet, now called C/1786 P1 Herschel and sometimes called "*the first lady's comet*," as defined by the novelist Fanny Burney. Its magnitude was 7.5, and the sky conditions were not optimal, so that they had to wait until the next night to confirm the discovery. The first observation of the comet with the naked eye occurred on August 17, and Charles Messier the following night observed with his telescope a long tail of 1.5°. The object quickly became easier to be observed, and on August 19 Caroline's brother described the comet as "considerably brighter than the globular cluster M3," with a magnitude between 5 and 6. Observations continued until October 26, and calculations indicated that the comet had reached perihelion on July 8 at a distance of only 0.41 astronomical units (about 61 million km).

It has to be stressed that, at the time, the discovery of a comet was one of the most important things for an observer. To better understand this point, it can be remembered that the famous catalog bearing the name of the astronomer Charles Messier was issued by this celebrated observer just to help him and his colleagues in their main activity as comet-hunters, so that they could not be wrongly identified with the object of their research.

Her excitement for this discovery shines in her notes:

August 1. – I have counted one hundred nebulæ to-day, and this evening I saw an object which I believe will prove to-morrow night to be a comet.

2nd. – To-day I calculated 150 nebulæ. I fear it will not be clear to-night. It has been raining throughout the whole day, but seems now to clear up a little.

1 o'clock. – The object of last night is a comet.

3rd. – I did not go to rest till I had wrote to Dr. Blagden and Mr. Aubert to announce the comet. After a few hours' sleep, I went in the afternoon to Dr. Lind, who, with Mr. Cavallo, accompanied me to Slough, with the intention of seeing the comet, but it was cloudy, and remained so all night.

And the aforementioned letter to Dr. Blagden is:

August 2, 1786.

SIR –

In consequence of the friendship which I know to exist between you and my brother, I venture to trouble you, in his absence, with the following imperfect account of a comet:–

The employment of writing down the observations when my brother uses the twenty-foot reflector does not often allow me time to look at the heavens, but as he is now on a visit to Germany, I have taken the opportunity to sweep in the neighbourhood of the sun in search of comets; and last night, the 1st of August, about 10 o'clock, I found an object very much resembling in colour and brightness the 27 nebula of the Connoissance des Temps, with the difference, however, of being round. I suspected it to be a comet; but a haziness coming on, it was not possible to satisfy myself as to its motion till this evening. I made several drawings of the stars in the field of view with it, and have enclosed a copy of them, with my observations annexed, that you may compare them together.

August 1, 1786, 9h 50′. Fig. 1. The object in the centre is like a star out of focus, while the rest are perfectly distinct, and I suspect it to be a comet.

10h 38′. Fig. 2. The suspected comet makes now a perfect isosceles triangle with the two stars a and b.

11h 8′. I think the situation of the comet is now as in Fig. 3, but it is so hazy that I cannot sufficiently see the small star b to be assured of the motion.

By the naked eye the comet is between the 54 and 53 Ursæ Majoris and the 14, 15, and 16 Comæ Berenices, and makes an obtuse triangle with them, the vertex of which is turned towards the south.

Aug. 2nd, 10h 9′. The comet is now, with respect to the stars a and b, situated as in Fig. 4, therefore the motion since last night is evident.

10h 30′. Another considerable star, c, may be taken into the field with it by placing a in the centre, when the comet and the other star will both appear in the circumference, as in Fig. 5.

These observations were made with a Newtonian sweeper of 27-inch focal length, and a power of about 20. The field of view is 2° 12′. I cannot find the stars a or c in any catalogue, but suppose they may easily be traced in the heavens, whence the situation of the comet, as it was last night at 10h 33′, may be pretty nearly ascertained.

You will do me the favour of communicating these observations to my brother's astronomical friends.

I have the honour to be,

Sir,

Your most obedient, humble servant,

CAROLINA HERSCHEL.

Mr. Aubert after seeing the comet, replied to her: *"You have immortalized your name"*.

This discovery therefore led some notoriety to Caroline, and she became the subject of several articles. Miss Burney, the writer and lady-in-waiting to the Queen, in 1787 described her as *"[. . .]very little, very gentle, very modest, and very ingenuous; and her manner are those of a person unhackneyed and unawed by the world, yet desirous to meet and return its smiles"* and Mrs. Papendick portrayed her as *"[. . .] by no means prepossessing, but an excellent, kind-hearted creature."*

21.4   Assistant to the King's Astronomer

In 1787 King George III recognized the activity of Carolina as an assistant of her brother William by assigning to her a salary of 50 pounds per year. Despite it being a moderate sum, still the fact is very important because it makes her the first woman in the UK, and the second in the world after Christine Kirch, whose professional work in astronomy became officially recognized and remunerated. The history of this recognition is not just a straightforward consequence of the discovery of a comet, as one might imagine, rather it is probably more complex, as pointed out by the historian of Astronomy Michael Hoskin.

Portrait of Caroline Herschel. From *The Scientific Papers of Sir William Herschel*

William in fact had known Mary Pitt, a widow who he desired to marry, but this required some time for negotiation both with her and with Caroline, who would have half to leave the house where she lived with her brother. At first he thought to give her an economical compensation which would have made it possible for Caroline to live on her own, but she firmly refused such an arrangement, which would have resembled a sort of charity. William therefore agreed to appeal to the king for a grant that included a salaried position for his sister. It is interesting to notice that, in his application letter, the astronomer advocated his requests by claiming that an assistant needed to operate the instrumentation would have cost two times the required salary.

The following year, eventually, William married Mary Pitt and Caroline's habits had to change: "*Initially Caroline was deeply affected by the marriage, and moved out to lodgings at Upton. She continued to support her brother's work and in making the daily walk to Observatory House, became a well-known figure. Often, with William resting after a long night of observation, the house was kept as quiet as possible during the day. Eventually the relationship between the two ladies - Mary and Caroline – warmed [...]*".

Caroline remembered with great sorrow the change of the relationship with her brother and also recalled her bitterness towards the sister-in-law. However, their relationship improved with time, to the point that, probably regretting what she had written, she later destroyed the pages of her diary about this period. Eventually this event made her independent from her brother, for whom before his marriage she was assisting like a housekeeper. By living in a house next to the one of the new couple, the two brothers could continue to enjoy their mutual collaboration, but on a more autonomous footing, from both the economic and the scientific point of view, actually favoring the flourishing of the sister's astronomical career, in which she especially excelled as a comet-hunter.

On December 21, 1788 Caroline discovered her second comet, which was called 35P/Herschel-Rigollet at about 1° south from the star Beta Lyrae. William Herschel described it as a considerably bright and irregularly shaped nebula, brighter in the center, five or six arcminutes in diameter. The chase continued until February 5, 1789, and its orbit, believed parabolic, reached perihelion on November 21, at a distance of 0.75 astronomical units. About 151 years later, on 28 July 1939, Roger Rigollet discovered a comet of eighth magnitude, and subsequent orbital calculations confirmed that it was the same comet discovered by the Anglo-German astronomer in 1788, hence the double name. Although not parabolic, its orbit is quite elliptical. Observed for the last time on January 16 1940 from the Lick Observatory, its next passage was foreseen by the end of the twenty-first century. When it was introduced and given its present cometary nomenclature, in early 1995, the comet 35/P Herschel-Rigollet was assigned the prefix that identifies it as the 35th periodic comet to be observed at perihelion.

Over a period of 10 years, Caroline's comet discoveries piled up to the impressive result of 8, a female record which was surpassed only in 1980 by another Caroline, she astronomer Carolyn Shoemaker. More precisely, only six comets can be officially ascribed to her, but for two more she was significantly involved in their identification. In 1790 in fact there were two (C/1790 A1 Herschel and C/1790 H1 Herschel) in January and April. The following year, exactly on December 15, 1791 it came the fifth one (C/1791 X1 Herschel), and almost 2 years later, on October 7, 1793, C/1793 S2 Messier which however had already been sighted without her knowing by Charles Messier, as it can be understood by the official denomination. The seventh and the eighth comets are the 2P/Encke and the C/1797 P1 Bouvard-Herschel, but while the latter has a quite normal history, with the exception of a contemporary discovery by our protagonist and by the astronomer Alexis Bouvard on August 14, 1797, the former has a much more complex and interesting history which deserves a separate telling.

21.5 Curiosity

The name of comet 2P/Encke means that, unlike the above ones, it is a periodic object, that is one with a closed and known orbit. It was sighted on the night of January 17, 1786, in the constellation of Aquarius, by the famous comet hunter

Pierre Mechain. At the time it appeared it was an object of magnitude 6.3, like the globular cluster M2. After having reported his discovery to Messier, both comet hunters, along with Jean Dominique Cassini, observed it again two nights later, on January 19, but because of its fast moving, the comet was no longer sighted and therefore, on the basis of only two observations, it was not possible to calculate its orbit. Almost 10 years passed before the comet could be sighted again by Caroline, on the night of November 7, 1795, as an object with a brightness which could be compared to that of M31. To that time no one knew that it was the same comet of Mechain, and therefore it was classified as a new comet whose discovery was attributed to the Herschel sister. During this passage it was then observed for 3 weeks by other astronomers like Johan Bode and Heinrich Olbers, and an orbit determination was also attempted, but it could only be concluded that it was not parabolic.

Another 10 years later the comet was rediscovered once again, on October 19, 1805 by Jean Louis Pons, Johann Sigismund Huth and Alexis Bouvard. This time, however, Johann Encke, a German astronomer who later became director of the Observatory of Berlin, announced that from the study of these observations the orbit had to be elliptical with a period of 12.1 years, not a very accurate estimation, but still much more correct than the calculations of other astronomers who were still trying to find a parabolic orbit.

The comet appeared again in 1818. Pons saw it on November 26, and it remained visible for about 7 weeks. Olbers then suggested that it had to be the same comet observed in 1786, 1795 and 1805, and Encke provided the mathematical proof that it was actually the same comet and that it had a period of 3.3 years. In 1819 Encke, thanks to his orbital solution, computed its previous passages by taking into account the perturbations caused by the known planets, with the exception of Uranus, and in 6 weeks he was able to confirm that the four comets were truly the same.

With the confirmation of its periodicity Encke could predict the next coming of the comet, with May 24, 1822 as the date of its passage at perihelion, and on June 2 Karl Rümker sighted the comet from a private observatory in Australia. This was the second comet after the Halley whose return was predicted, and for this reason it took the name from Encke.

The comet huntress Caroline, with eight trophies, at the time took the third place after two famous French astronomers: Messier with 14 and Méchain with 10; she enjoyed the chase, but took no interest in their scientific significance. All the documents relating to Caroline's comets were found after her death neatly assorted in a packet labelled: *"Bills and Receipts of my Comets"*.

Another curious fact regards the "family enterprise" of the Herschels or, to say it better, of William. His activity was in part funded by building and selling telescopes. This was a completely artisan activity, so the instructions for use and assembly included in the telescopes were not printed, but written accurately by hand by Caroline.

21.6 As They Said of Her

In 1791 Caroline used a telescope that she took with her when she retired to Hanover, and now the optics survive in the *Historisches Museum* in the same city. Nevil Maskelyne, the Astronomer Royal, wrote about it: "*[...] She [Caroline Herschel] shewed me her 5 feet Newtonian telescope made for her by her brother for sweeping the heavens. It has an aperture of 9 inches, but magnifies only from 25 to 30 times, & takes in a field of 1° 49' being designed to shew objects very bright for the better discovering any new visitor to our system, that is Comets, or any undiscovered nebulae. It is a very powerful instrument, & shews objects very well. It is mounted upon an upright axis, or spindle, and turns round by only pushing or pulling the telescope; it is moved easily in altitude by strings in the manner Newtonian telescopes have been used formerly. The height of the eye-glass is altered but little in sweeping from the horizon to the zenith. This she does and down again in 6 or 8 minutes, & then moves the telescope a little forward in azimuth, & sweeps another portion of the heavens in like manner. She will thus sweep a quarter of the heavens in one night. The Dr [William Herschel] has given her written instructions how to proceed, and she knows all the nebulae [listed by Messier] at sight, which he esteems necessary to distinguish new Comets that may appear from them. Thus you see, wherever she sweeps in fine weather nothing can escape her.*"

Caroline was the first woman to officially discover a comet and Miss Fanny Burney, the novelist, called "*her eccentric vocation*"; William showed Caroline's first comet to the Royal Family, but Caroline was not there because she wasn't the King's astronomer. But also present was Miss Burney who said: "*The comet was very small, and had nothing grand or striking in its appearance; but it is the first lady's comet, and I was very desirous to see it.*"

A leading English amateur, Francis Wollaston, said: "*Miss Herschel whom I put first as a sister astronomer*", one of innumerable visitors to the Herschel home, Professor Karl Seyffer of Göttingen called her the "*most noble and worthy priestess of the new heavens*" and Lalande sent her a letter of thanks for news of the fourth comet to "*Madamoiselle Caroline Herschel, astronome célèbre (famous astronomer)*".

As mentioned before, Caroline had refused William's money offer that would make her independent on the event of his marriage. It was because of this that he asked the King for a salary for his assistant: "*You know Sir, that observations with this great instrument cannot be made without four persons: the Astronomer, the assistant, and two workmen for the motions. Now, my good industrious sister has hitherto supplied the place of assistant, and intends to continue to do that work. She does it indeed so much better, to my liking, than any other person I could have , that I should be very sorry ever to lose her from that office. Perhaps our gracious Queen, by way of encouraging a female astronomer, might be enduced to allow her a small annual bounty, such as 50 or 60 pounds, which would make her easy for life, so that, if anything should happen to me she would not have the anxiety upon her mind of being left unprovided for. She has often formed a wish but never had the resolution of causing an application to be made to her Majesty for this purpose; nor could I have been prevailed upon to mention it now, were it not for her evident use in the observations that are to be made with the 40 feet reflector, and the unavoidable increase or the annual expences which, if my Sister were to decline that office would probably amount to nearly one hundred pounds more for an assistant."*

Her nephew John wrote: *"I learned fully to appreciate the skill, diligence, and accuracy which that indefatigable lady brought to bear on a task which only the most boundless devotion could have induced her to undertake and enable her to accomplish."*

William Herschel's 40-foot telescope, 1789

21.7 Awards

Meanwhile Caroline's scientific work continued, and in addition to her activity as a comet sweeper she embarked alone on a new project whose aim was to check and correct the star catalog produced by John Flamsteed. In 1798 Caroline presented to the Royal Society the *Index to Flansteed's Observations of the Fixed Stars*, together with a list of 560 other stars that had been neglected. This publication marks the temporary end of her work as a researcher, which started again 25 years later, after the death of her brother William. In this period she devoted herself to the education of the son of William and Mary, John Herschel, who was born in 1792, and who spent long periods with her during the holidays. During her University studies John was at Cambridge, where he excelled as a mathematician. The Royal Society recognized very soon his merits, choosing him to join the astronomical research of his father, and awarding the young scientist with the Copley Medal in 1821 for his achievements. In these years Caroline's fame earned her a lot of important attention. She was the guest of the Astronomer Royal Nevil Maskelyne at the Royal Observatory in 1799 and of members of the royal family on several occasions in 1816, 1817 and 1818, receiving awards for her scientific merits.

In 1822, upon the death of her beloved brother William, she decided to return to Hanover. It was surely with sorrow that she had resolved to leave England, but in this way she was able to pursue some research projects which helped her nephew John. Moreover, such research was not done as an assistant, but as an independent researcher. She therefore produced a catalog of 2500 nebulae for which the Royal Astronomical Society in February 1828 awarded her its gold medal: *"That a Gold Medal of this Society be given to Miss Caroline Herschel for her recent reduction, to January, 1800, of the Nebulae discovered by her illustrious brother, which may be considered as the completion of series of exertions probably unparalleled either in magnitude or importance in the annuals of astronomical labour."* This award was not to be received by another woman until 1996.

In Hanover she became a celebrity as well, and many scientists, such as the mathematician Gauss, payed her a visit. Saying that the most appreciated one was that of her grandson, however, would be an easy win. Indeed in 1832, when she was eighty-two, John wrote of her: *"She runs about the town with me, and skips up her two flights of stairs. In the morning, till eleven or twelve, she is dull and weary, but as the day advances she gains life, and is quite 'fresh and funny' at ten p.m., and sings old rhymes, nay, even dances."*

In 1835 Caroline Herschel and another scientist, Mary Somerville, were elected honorary members of the Royal Society, becoming the first women to receive this prestigious title.

It would be tempting to suggest that the two main female astronomers of this period knew each other personally, since they lived close to each other for more than 5 years, however there is no record of such fact. What can be stated, instead, is that Caroline was admired by her younger colleague, who also sent a letter and a copy of one of her books.

She was also elected a member of the Royal Irish Academy in 1838 and for her ninetieth birthday the celebrated astronomer received a letter saying: *"His Majesty the King of Prussia, in recognition of the valuable service rendered to astronomy by you, as the fellow worker of your immortal brother, wishes to convey to you in his name the Large Gold Medal for science."* And during the celebrations for her birthday she *"[. . .] entertained the crown prince and princess with great animation for two hours, even singing to them a composition of her brother William."*

On January 9, 1848 at the age of 97, merely 2 months before her 98th birthday, Carolina died in Hanover. She herself wrote the beginning of her epitaph, which reads:

Here rests the earthly exterior of
   CAROLINE HERSCHEL,
   Born at Hanover, March 16, 1750.
   Died January 9, 1848.
   The eyes of Her who is glorified were here below turned to the starry Heavens. Her own Discoveries of Comets, and her participation in the immortal Labours of her Brother, William Herschel, bear witness of this to future ages.
   The Royal Irish Academy of Dublin, and the Royal Astronomical Society of London enrolled Her name among their Members.
   At the age of 97 years 10 months she fell asleep in calm rest, and in the full possession of her faculties, following into a better Life her Father, Isaac Herschel, who lived to the age of 60 years 2 months 17 days, and lies buried not far off, since the 29th of March, 1767.

Her fame continues like the tributes to her: an asteroid (281) discovered on October 31, 1888 by J. Palisa in Vienna, in 1889 was named Lucretia after her. In 1935 a Lunar crater with a diameter of 13 km was called Carolina Herschel (34.5N, 31.2W). Such are the tributes to a woman of small personal ambitions that did not like the praises of strangers, but cherished those of her brother William.

Chapter 22
Margaret Bryan (1760?–1816)

... even the learned and more difficult sciences are ... beginning to be successfully cultivated by the extraordinary and elegant talents of the female writers of the present day. (Charles Hutton)

She looks a fine and handsome lady with her two young daughters. Margaret Bryan in this miniature appears as if she has just stopped working, and we can also notice in the room and over the desk several astronomical objects like a telescope, a planetary and an astrolabe.

© Springer International Publishing Switzerland 2016

G. Bernardi, *The Unforgotten Sisters*, Springer Praxis Books in Popular Astronomy, DOI 10.1007/978-3-319-26127-0_22

Portrait of Margaret Bryan and her daughters

Was she an astronomer? Or is she teaching her daughters? Actually, we don't know where and when she was born or died exactly, but we know that Margaret Bryan was a British natural philosopher, educator and populariser of science who published three scientific textbooks which became standards in the British educational institution during the final years of the eighteenth and early nineteenth century. What we know of her life, in fact, comes mainly from her working activity, and very little about her private life has survived to our days. Margaret was a talented schoolmistress, probably born some time before about 1760 and wife of a Mr. Bryan, but we don't know anything about him.

22.1   Works

Margaret Bryan's school was located in London or in its neighbourhoods, but apparently moved several times since three of its addresses or locations have been handed down to our days: one is at Blackheath, another at 27 Lower Cadogan

Place near Hyde Park Corner, and the last at Margate. We know that after August 1797 she was published in London by Leigh and Sotheby and G. Kearsley, her first book: *A Compendious System of Astronomy*, dedicated to the pupils of her school. As it was customary at the time, the complete title was almost a comprehensive summary and gives us a clear view of its content:

- *A Compendious System of Astronomy.* Title page Title: *A Compendious System of Astronomy, in a course of familiar lectures: in which the principles of that science are clearly elucidated, so as to be intelligible to those who have not studied the mathematics. Also trigonometrical and celestial problems, with a key to the ephemeris, and a vocabulary of the terms of science used in the lectures: which latter are explained agreeably to their application in them.*

Indeed, the book was composed of ten lectures on astronomy, optics, Newton's laws, gravity, planetary orbits, motions of the Moon, eclipses, transits, fixed stars and the universe. It was written in a conversational style and with simple and clear language suited to her young lady pupils, and was also illustrated with beautiful and detailed drawings and diagrams, apparently drawn by the author herself. The book included also the latest findings of that time such as the planet Uranus, discovered by William Herschel in 1780, and still called Georgium Sidus, in honour of Herschel's patron King George III. In the first edition Margaret recorded also the discovery of two moons around Uranus and in the second edition of 1799 she increased this number to six. An introduction on elementary plane geometry and trigonometry, and a set of problems with solutions completed the work.

The next year, in April she defended herself in the *Critical Review* against what she felt to be damaging criticism that appeared in the same journal on her *Compendious System*. The *Compendious System* had been published by subscription, a system to collect the money needed to cover the publication costs by private contributions, and which we would now call "crowdfunding". For this reason the book, after the preface, opens with the list of subscribers in alphabetical order. The same method was adopted to fund the publication, in 1806, of another Physics book, the *Lectures on Natural Philosophy*, which contained thirteen lectures on acoustics, hydrostatics, mechanics, pneumatics, magnetism and electricity with a portrait of the author, engraved by Heath, after a painting by T. Kearsley. The last few chapters are dedicated to astronomy: optics, lenses, mirrors, telescopes, microscopes, the spectrum and spectroscope with drawn diagrams, numerous tables and simple problems. There is also a notice in it that *"Mrs Bryan educates young ladies at Bryan House, Blackheath"* which probably indicates that the seat of the school was her private home.

Finally, a third and last book is entitled *A Comprehensive Astronomical and Geographical Class Book for the use of Schools and Private Families,* described as a thin octavo in the list of works of 1816 in 'The British Review, and London Critical Journal'. After this year it is not known exactly when or where she moved.

22.2   Curious Fact

"*Mrs. Bryan receives young ladies for the purpose of education.*" This is reported as advertisement at the end of the 1797 edition of the *Compendious System*. Does this mean that she had not officially established her school? Margaret Bryan's school, however, had one remarkable difference from its contemporaries because it was an academy where girls could learn mathematics and science.

In 1799 the second edition of the book displayed a list of 400 subscribers who financed the publication which now included distinguished scientists and members of high society, a fact suggesting that she could enjoy good connections, maybe as a member of that same circle. For example Charles Hutton, professor of mathematics at the Royal Military Academy at Woolwich introduced her to the astronomer William Herschel. At Slough in his observatory, Mrs. Bryan once visited him and his sister Caroline, about whom it will be extensively written later on.

Margaret was also an observer since in 1811, in a letter to William Herschel, she described her efforts to observe a comet of that year by Jean Louis Pons.

In the Drawing Room, inside the Herschel Museum of Astronomy in Bath, there is a painting by miniaturist Samuel Shelley showing Margaret Bryan and her two daughters. This is the cover of the book titled *A Compendius System of Astronomy in a Course of Familiar Lectures*. She dedicated the book to her pupils and this miniature and the copy of the second edition of this publication were purchased through a grant aid from The PRISM Fund and the Royal Astronomical Society.

22.3   As They Said of Her

The Dictionary of National Biography does not tell us much about Mrs. Bryan, describing her only as "*a beautiful and talented schoolmistress.*"

The aforementioned Charles Hutton endorsed the work of the 'beautiful' Mrs. Bryan in a letter dated Jan 6, 1797 where she stated that "*[. . .] even the learned and more difficult sciences are thus beginning to be successfully cultivated by the extraordinary and elegant talents of the female writers of the present day.*" We know of this letter since she included the full text at the end of the Preface.

Chapter 23
Wang Zhenyi (1768–1797)

Actually, it's definitely because of the moon.

Wang Zhenyi was a Chinese scientist who lived during the Qing dynasty. Her name in the simplified Chinese is 王贞仪, in traditional Chinese 王貞儀, and pinyin Wáng Zhēnyí.

In 1768 Wang Zhenyi was born into a prominent family from the province of Anhui. Her grandfather, Wang Zhefu (王者 辅), was a former governor of the Fengchen county and Xuanhua District. He was an intelligent man, with lively cultural interests who owned a library containing more than 70 volumes. These books became a familiar presence since the earliest years of his granddaughter, and acted as a stimulus for the young Wang Zhenyi, who soon learned to read in order to know their content. Also her family had a positive influence in the broadening of her knowledge: her father, Wang Xichen, had failed the imperial examination required to become a civil servant, so he devoted himself to medicine by recording his findings in a collection of four volumes called *Yifang yanchao* or "Collection of Medical Prescriptions." He thus became her first teacher of medicine, geography and mathematics. Her grandmother Dong taught poetry, while her grandfather, who however died in 1782, introduced her to astronomy. For his funeral the family went to Jiling, near the Great Wall, and remained there for 5 years. Here the young Zhenyi completed her training, which was not limited to the development of her intellectual skills. Actually, she acquired equestrian skills, archery and martial arts from the wife of a Mongol general called Aa. When she reached the age of 16 she traveled with her father to the south of the Yangtze River, visiting the provinces of Shaanxi, Hubei and Guangdong and at eighteen she entered in friendship with female scholars of Jiangning, today's Nanjing, through her poetry. In addition to this, as self-taught she pursued her studies of astronomy and mathematics. At the age of 25 she married Zhan Mei Xuan Cheng, of the Anhui province. Her marriage was very happy, and she became even better known as a poet, an astronomer and a mathematician, teaching also a few male students, but at the age of 29 she died childless.

© Springer International Publishing Switzerland 2016
G. Bernardi, *The Unforgotten Sisters*, Springer Praxis Books in Popular Astronomy, DOI 10.1007/978-3-319-26127-0_23

23.1   Works and Experiments

The life of Wang Zhenyi was very short, nonetheless she left several mathematical and astronomical contributions. As a self-taught scholar herself, the difficulties she had met during her mathematical studies had to be quite hard if she once said: *"There were times that I had to put down the pen and sighed. But I love the subject, I do not give up."* She thus always understood very well the importance of having clear and accessible scientific texts, and wrote her works with this intention in mind. Among the others, she admired the "Principles of Calculation" of the famous mathematician Mei Wending (1633–1721), of which she became a great connoisseur. She then rewrote it in a simple and accessible language under the title "The Musts of Calculation." Another trick used by Wang Zhenyi to make math learning easier for beginners was to simplify the way to perform multiplications and divisions. This culminated in a book on such subject that she wrote at twenty-four and entitled "The Simple Calculation Principles." Still in the field of mathematics, and especially in trigonometry, Zhenyi wrote an article entitled "The Explanation of the Pythagorean Theorem and Trigonometry."

As an astronomer, she was able to describe the celestial phenomena in a simple way. For example, in her work "Dispute on the Precession of Equinoxes" she explains and demonstrates how the equinoxes move and shows how to calculate their movement. Other articles followed, like the "Dispute of Longitude and Stars" and "The Explanation of a Lunar Eclipse" where she analyzed the motions of the Moon and described the phenomena of Lunar and Solar eclipses. Her approach to the study was not limited to the understanding or the simplification of the texts and of the researches of other astronomers, but she was also able to pursue her own personal research.

A replica of Ming Dynasty's Armillary in the courtyard of Beijing Ancient Observatory

One example is her famous experiment on Lunar eclipses in order to explain why they occur. At a time when these phenomena were interpreted by most of the population as a sign of the anger of the gods, she wrote in one of her books: "*Actually, it's definitely because of the moon*", also developing what today might be called an exhibit. It was made up by the round table of the garden pavilion, a crystal lamp supported by a cable hanging from the rafters of the structure and that represented the Sun, and a round mirror that represented the Moon. Appropriately moving objects, for example, she was able to show that Lunar eclipses occur when the Moon passes into Earth's shadow.

Book

- The Simple Principles of Calculation.

Articles
"Dispute of the Procession of the Equinoxes."
"Dispute of Longitude and Stars."
"The Explanation of a Lunar Eclipse."
"The Explanation of a Solar Eclipse."
"The Explanation of the Pythagorean Theorem and Trigonometry."

23.2 What She Wrote, as They Said for Her and What She Said

In addition to scientific research, Wang Zhenyi, also devoted herself to poetry which she eventually collected in thirteen volumes of "Ci" or poetry. These add to prose, various prefaces and postscripts written for other works. The famous scholar Yuan Mei said that her works "*had the flavor of a great pen, not of a female poet.*" The poetry of Wang Zhenyi was known for the absence of flowery words common to the feminine style. The themes in her poems related to the classics and history. The numerous trips she made with her father were a great inspiration, giving her the inspiration to describe places, the lives of ordinary people and of working women, or the contrast between rich and poor. Here are some examples:

"Transiting Tong Pass"

So important is the doorway,
occupying the throat of the mountain
Looking down from the heaven,
The sun sees Yellow river streaming.

"Climbing Tai Mountain"

Clouds overcast the hills,
The sun bathes in the sea.

"A Poem of Eight Lines"

Village is empty of cooking smoke,
Rich families let grains stored decay;
In wormwood strewed pitiful starved bodies,
Greedy officials yet push farm levying.

Wang Zhenyi believed in equality and equal opportunity for both men and women. She wrote in one of her poems:

It's made to believe,
Women are the same as Men;
Are you not convinced,
Daughters can also be heroic?

In this regard, she publicly expressed her opinion against the common values of her contemporaries who confined the admissible female activities within a very limited range: "*When talking about learning and sciences, people thought of no women,*" and "*women should only do cooking and sewing, and that they should not be bothered about writing articles for publication, studying history, composing poetry or doing calligraphy.*" [Men and women] "*are all people, who have the same reason for studying.*"

In fact her situation was an unusual one for a woman, as she wrote: "*I have traveled ten thousand li and read ten thousand volumes. Bold is attempt to surpass men.*"

23.3   Awards

The International Astronomical Union in 1994 named Wang Zhenyi a crater on Venus: latitude 13.2, longitude 217.8 and diameter km 23.4.

Chapter 24
Marie-Jeanne Amélie Harlay Lefrancais de Lalande (1768–1832)

*My niece helps her husband in his observations and draws
conclusions by calculation. . . (Joseph Jérôme Lalande)*

Marie-Jeanne Amélie Harlay was born in Paris in 1768 and married at twenty
Michel Jean Jérôme Lefrancais de Lalande (1766–1839), grandson of the famous
astronomer Joseph-Jérôme Lalande, who was an astronomer like his uncle. They
had a daughter and her celebrated great-uncle wrote: *"This son of astronomy was
born on January 20, 1790, a day where we saw in Paris for the first time the comet
discovered by Miss Caroline Herschel; the child was then given the name of
Carolina; his godfather was Delambre."*

Marie-Jeanne Amélie helped her husband with his astronomical observations
and performed the mathematical calculations required for their interpretation. She
became so expert that she helped the astronomer Cassini in 1791, then director of
the Paris Observatory, in his observations at the College of France.

© Springer International Publishing Switzerland 2016
G. Bernardi, *The Unforgotten Sisters*, Springer Praxis Books in Popular Astronomy,
DOI 10.1007/978-3-319-26127-0_24

The Paris Observatory represented by Louis Figuier at about 1870

24.1 Works

- *Tables horaires* de *l'Abrégé de navigation* de Joseph Jérôme Lalande, Paris.
- *Histoire celeste francaise*, de Joseph Jérôme Lalande, Paris 1801.

In 1793 Marie-Jeanne calculated and published the *Table horaires pour la marine*—Time tables for the navy—contained in the *Abrégé de navigation* (navigation extract) of Joseph-Jérôme Lalande. These tables occupied three hundred pages and was defined a *"huge job for her age and her sex"* by her uncle. Such tables were common at the time, and were used by sailors to determine the time at sea by calculating the altitude of the Sun and the stars. These tables earned their author a medal of Lycée des Arts, which was awarded to scholars and distinguished artists. "*I believe,* - writes Lalande - *that what women are missing is just the education opportunities and examples which they can emulate; we see them standing out enough, despite the obstacles of the education and of the prejudices, to believe that they have just as much talent as most of the men who get a reputation in Science.*"

Madame de Lalande collaborated in preparation of the *Histoire céleste francaise* of Lalande, the largest and most complete star catalog of the time which was published in 1801 and included positions and magnitudes of more than 47,000 stars.

The Muse Urania by Johann Heinrich Tischbein the Elder, 1782

The amount of work implied by this publication can be understood by the fact that, just considering the computational part, at least 36 operations were needed for each star. As a collaborator of Lalande and of those works, her name became very well known throughout Europe. The famous physicist Gauss became acquainted with her before that of Sophie Germain.

She died in 1832 at the age of 64.

24.2   As They Said of Her

Joseph-Jérôme Lalande, director of the Astronomical Observatory of Paris between 1795 and 1800, published a note of appreciation for her niece Marie-Jeanne:

My niece helps her husband in his observations and draws conclusions by calculation. She has reduced the observations of ten thousand stars and prepared a work of three hundred pages of time tables, a gigantic work for her age and her sex, which are published in my Abrégé de navigation. She is one of the few women who have written books of science. She published tables to calculate the time at sea based on the altitude of the Sun and of the stars. These tables were published in 1791 by order of the National Assembly [. . .] in 1799 she published a catalog of ten thousand stars, reduced and calculated.

24.3   Curious Facts

She gave her daughter the name of Caroline after the great Anglo-German astronomer Caroline Lucretia Herschel. Caroline Lalande was born in Paris on 20 January 1790, the same day when there was spotted in the city one of the many comets discovered by her famous astronomer forerunner.

Chapter 25
Mary Fairfax-Somerville (1780–1872)

The Queen of Nineteenth Century Science

On December 26, 1780 little Mary Fairfax was born in Roxburghshire, Scotland, to be more precise in the city of Jedburgh. The customs of the time would have assigned to her a quite conventional future, but her fate took a different direction. Until about 10 years of age, as it was customary for the girls of the time, she received an approximate education, although she had been sent for 1 year by her father—a high-ranking navy officer who was often at sea—at an expensive boarding school. The young girl, however, later began a self-teaching education which included Latin and mathematics. In part the cause of her interest can be ascribed to her uncle, Dr. Thomas Somerville, who eventually became her father-in-law in a second wedding. Dr. Somerville often told Mary stories of the great scientists of the ancient world, which inspired her young mind, and also helped his niece with Latin. The curiosity of this future female scientist, however, was not limited to past history, but it was geared to understanding the strange symbols of algebra. Yet in order to deepen her knowledge of this area of mathematics she needed appropriate textbooks, so she devised a stratagem. Since a woman could not directly buy such books, she asked the tutor of her younger brother to find her a copy of Euclid's *Elements*, and once she got it she began to study the mathematical problems presented in women's magazines. At home, however, this talent was not highly regarded: "[...] *we shall have Mary in a strait jacket one of these days. There was X., who went raving mad about the longitude!*" This is what Mary's father once said, and the family was unanimous in discouraging her strange passion, to the point that her parents attempted to stop these readings by removing the candles. But Mary did not give up and memorized the problems that she could work mentally to the solutions. Years passed, and in 1804 Samuel Grieg, captain of the Russian Navy, became her first husband. As she recalled: "*My husband had taken me to his bachelor's house in London, which was exceedingly small and ill ventilated. I had a key of the neighbouring square, where I used to walk. I was alone the whole of the day, so I continued my mathematical and other pursuits, but under great disadvantages; for although my husband did not prevent me from studying, I met with no sympathy whatever from him, as he had a very low opinion*

© Springer International Publishing Switzerland 2016
G. Bernardi, *The Unforgotten Sisters*, Springer Praxis Books in Popular Astronomy,
DOI 10.1007/978-3-319-26127-0_25

of the capacity of my sex, and had neither knowledge of, nor interest in, science of any kind." After 2 years of marriage and two children Mary remained a widow, and she returned to Scotland with them. With a modestly comfortable inheritance, she was now able to devote herself to her scientific interests in mathematics, winning in 1811 a prize for a solution of a Diophantine equation, and beginning to approach the study of Newton's *Principia*.

25.1   Works

In 1812 she married again, this time with her cousin William Somerville, a physician who returned to Scotland after nearly 15 years abroad in the army medical department. The two shared an interest in natural sciences and four children were born to them. The new husband significantly encouraged and supported her scientific activity.

Portrait of Mary Fairfax, Mrs William Somerville, 1834

As a member of the Royal Society he gave her the opportunity to access the library of the institution, and to be presented to several scientists. When she became a writer of popular science it was he who edited and copied his manuscripts, who compiled the bibliographies, and who managed the correspondence with scientists and publishers. Together with her husband she studied geology and, at the age of 33, Mary started to study Greek, botany, higher mathematics, physics, meteorology and astronomy. In 1816 the Somervilles moved from Edinburgh to London, an environment rich in scientists and professionals. Just to give some examples,

William Wollaston showed her the prism he had used to discover the solar spectrum just within hours from the discovery, while Thomas Young explained his astronomical method for dating the Egyptian papyri, and Sir James South taught her to observe binary systems. She was allowed to see the computing machines of Charles Babbage and became an intellectual counselor of the young Ada Lovelace, daughter of Lord Byron, collaborator of Babbage and first programmer in history. Indeed a programming language, Ada, used primarily in the military, bears her name.

Watercolor portrait of Ada King, Countess of Lovelace (Ada Lovelace), 1840

Their association with the scientific world was not limited to its local or English branch. The Somervilles paid visits to the Herschels' observatory at Slough and met in Paris and in Switzerland the greatest scientists of the time. Such relations gave her the chance to receive books and articles on different subjects, and invitations to conferences and lectures. A perfect situation for the development of her activity and to write directly about science. For example, in 1838, when she started her journey in Italy and arrived at Turin, Mary was paid homage by Giovanni Plana, then director of the local Observatory, of his work *Théorie du mouvement de la lune* (Theory of the motion of the Moon). Several experiments were prepared, and the scientists responded to all her questions.

All this stunning success in science writing, however, marked the range of her activity and in a certain sense evidenced a kind of limitation. Indeed Mary's first work, which was read by her husband to the Royal Society since women were not allowed to attend the meetings of the Academy, was titled *"On the Magnetizing*

Power of the More Refrangible Solar Rays." It was a truly scientific work which was eventually published in 1826 in the *Philosophical Transactions of the Royal Society of London,* describing the results of an experiment which seemed to show that the blue, green and violet part of the Solar spectrum had a magnetic effect, in the sense that they could induce magnetism of non-magnetic metal objects like clock springs and sewing needles. These conclusions were accepted at first, stimulating further scientists, for example Herschel and some researchers in Vienna. They successfully repeated the work, which was accepted as valid for a number of years, but were later refuted, because other investigators showed that the effect was not due to the sun's violet rays. Judging from her own words, it seems that in spite of her research activity she did not consider herself fitted for "creative" science, believing that this was a distinct character of her sex: *"In the climax of my great success, the approbation of some of the first scientific men of the age and of the public in general I was highly gratified, but much less elated than might have been expected, for although I had recorded in a clear point of view some of the most refined and difficult analytical processes and astronomical discoveries, I was conscious that I had never made a discovery myself, that I had no originality. I have perseverance and intelligence but no genius, that spark from heaven is not granted to the sex, we are of the earth, earthy, whether higher powers may be allotted to us in another state of existence God knows, original genius in science is hope-less in this.*" It may seem strange to read such words from one of the most gifted female scholars of her time, but we have to consider that they were surely encouraged by the social attitude of her time which granted her such a high praise and success. Women interested in science could study physics and astronomy, and indulge in a "descriptive science" by showing and explaining the findings of their male colleagues, but it was implicitly regarded as inappropriate or outside of their capabilities to conduct experiments or original research.

Indeed, while it is difficult to establish if or to what degree certain limitations on the experimental and autonomous research were self-imposed, what can be established with certainty are Mary's great skills in the analysis and evaluation of scientific works and her ability to explain complex theories in a simple way without distorting them. An example in this sense is her translation and commentary of the *Mécanique Céleste* of the astronomer Pierre-Simon de Laplace, undertaken in 1827 at the express request of Lord Brougham on behalf of the Society for the Diffusion of Useful Knowledge. In this monumental work, the French scientist interpreted the astronomical observations of the Solar System objects, like the comets, the planets and the satellites, using Newton's theory of gravitation, proving that the Solar System is a stable and self-regulated mechanism. Such was its complexity that in 1808 it was told that no more than a dozen British mathematicians were able to read it, and Laplace was aware of the difficulty of this work. It is said that during a dinner with the Somervilles in Paris the mathematician, knowing nothing of his female guest's first marriage, commented: *"I write books that no one can read. Only two women have ever read the 'Mécanique Céleste'; both are Scotch women: Mrs. Grieg* [author's note: the first married name of Mary] *and yourself.*" In other sources the anecdote is reported in a slightly different way which, however, does not change the essence of its meaning: *"There have been only three women who have understood me. These are yourself, Mrs Somerville, Caroline Herschel and a Mrs. Greig of whom I know nothing."*

Mary began to work on this massive project with the understanding that, should they have found it unacceptable, they would have burned it. She committed herself to this job for 4 years, during which she contemporarily had to attend the social life and the education of their daughters because, as she wrote in her autobiography: *"A man can always command his time under the plea of business, a woman is not allowed any such excuse."*

The *Mechanism of the Heavens* was published in 1831 and, as mentioned above, it was more than a translation from French to English. Mary in fact presented in the introduction the mathematical concepts necessary for the understanding of the text, provided a historical profile of the subject, and she also completed the work with her own illustrations and demonstrations. It was a success not only among the common people, but also in the academic world, since it became a classic text of higher mathematics and astronomy until the end of the century adopted at the Cambridge, a university at which a woman would have not been able to study.

The second book, published in 1834 under the title *On the Connections of the Physical Sciences*, was a descriptive work that emphasized the growing interdependence among the various sciences, and in which more than a third of the work was devoted to her favorite subject: astronomy. It was a great success and was revised and reprinted ten times with the inclusion of the most recent discoveries. In the sixth and seventh year, Mary wrote: *"Those [tables of motion] of Uranus however are already defective, probably because the discovery of that planet in 1781 is too recent to admit of much precision in the determination of its motions, or that possibly it may be subject to disturbances from some unseen planet revolving about the Sun beyond the present boundaries of our system. If after the lapse of years, the tables formed from a combination of numerous observations should be still inadequate to represent the motions of Uranus, the discrepancies may reveal the existence, nay, even the mass and orbit of a body placed forever beyond the sphere of vision."* Confirming her assumption, the eighth edition of 1848 announced that John Adams and Urbain Leverrier, based on her observation, calculated the orbit of Neptune: *"[...] spent time with Airy and Adams [sic] the latter said Mr. S. that a consideration of my Phys. Sci. Has made up his mind to calculate the orbit of Neptune, had I been brilliant or original I could do it myself (a proof that the originality of the discovery is not given to women)."*

As it appears evident, in all her works Mary Somerville tried to describe and explain the current state of science in terms which were understandable to an educated reader. To this aim she always highlighted the experimental results with an accurate scientific vocabulary and, in case of disputes, by presenting the different positions and eliminating those discredited from the subsequent editions. The resulting impression is therefore that, with her characteristic style, she was particularly gifted in the art of science popularization by pulling people close to science without distorting its nature.

25.2   Curious Facts

During her life Mary Somerville received many awards, at home and abroad, but she was also attacked in several occasions. In 1835, sharing such exception with Caroline Herschel in an era when women had limited legal rights, she was granted an annual civil pension of 200 pounds, later increased to 300, which can be regarded as a salary for her scientific research. As remembered in the previous chapter, she also shared with her older colleague the honor of being the first female members of the Royal Astronomical Society, but she was also given honorary memberships of many other scientific societies.

At the time Carolina was 85 years old and Mary 55 and she send her a copy of her book accompanied by compliments:

ROYAL HOSPITAL, CHELSEA,
>April 16, 1835.
>DEAR MADAM, –
>I have sincere pleasure in availing myself of the opportunity of writing to you which the Astronomical Society of London has afforded me, by placing my name in the number of Honorary Members, and greatly adding to the value of that distinction by associating my name with yours, to which I have looked up with so much admiration.
>My object in writing is to request that you will accept of a copy of my book on the Connexion of the Physical Sciences, which is offered with great deference, having been written for a very different class of readers.
>I am proud of the friendship of your nephew, the worthy son of such a father, who is succeeding so well in his glorious undertaking at the Cape. I have seen a letter of the 27th January, when they were all well and prospering.
>I remain, dear Madam,
>With sincere esteem,
>Very truly yours,
>MARY SOMERVILLE

The two however never had the chance to need each other, since the only time Mary paid a visit to the Herschel observatory but Caroline was not there.

Moreover, the Royal Society of London then erected a bronze bust in the Academy, but quite interestingly she was never able to see it since, as a woman, she had no access to its premises. On the other side, after the publication of the Mechanism of Heavens she was accused of being a woman without God in the House of Commons, and the same Parliament denounced her again, together with influential members of the Church, when in her book *Physical Geography*, published in 1848, she accepted the theory of the antiquity of the Earth. In this regard, probably judging it a too delicate issue to be discussed, and although she knew and admired his work, she avoided altogether the evolutionary theory by Charles Darwin.

Arago in 1836 read extracts of a letter by Somerville to a meeting of the *Acadé mie des Science* (Academy of Science) where she sought to determine through experiments, whether the "chemical" sun's ray displayed activity analogous to light rays and "calorific" ray passing through various solid media. Mary demonstrated that rock salt transmitted the greatest number of chemical rays, using paper coated with silver chloride and prepared with Michael Faraday's advice, placed on various solids and exposed them all to the sun, but neither Mary nor Arago sensed that her method was a kind of primitive photography.

In 1838 the Somervilles moved to Italy, never to return, because of the worsening health of her husband, but her scientific activity never stopped. In her life she wrote several works:

- The Mechanism of the Heavens, 1831;
- The Connection of the Physical Sciences, *9th Edition*, 1858;
- Physical Geography, *6th Edition*, 1870;
- Molecular and Microscopic Science, 1869;

some of which will have influence on renown scientists like James Maxwell, including the aforementioned *Physical Geography* which will become her most famous text, adopted until the early twentieth century in schools and universities.

The *"Queen of the Nineteenth Century Science"* was engaged not only in science, but also with feminism, and she gave the former a great deal of importance as a way to improve the female condition in the society. In such context it is interesting to notice that, when Mary's eldest daughter died at the age of ten, she felt guilty for having encouraged her intellectual exercise, while in later years she wrote: *"Age has not abated my zeal for the emancipation of my sex from the unreasonable prejudice, too prevalent in Great Britain, against a literary and scientific education for women. Madame Emma Chenu, who had received the degree of Master of Arts from the Faculty of Sciences of the University in Paris, has more recently received the diploma of Licentiate in Mathematical Sciences from the same illustrious Society, after a successful examination in algebra, trigonometry, analytical geometry, the differential and integral calculus, and astronomy. A Russian lady has also taken a degree; and a lady of my acquaintance has received a gold medal from the same Institution. I joined in a petition to the Senate of London University, praying that degrees might be granted to women; but it was rejected. I have also frequently signed petitions to Parliament for the Female Suffrage, and have the honour now to be a member of the General Committee for Woman Suffrage in London."*

Quite properly, the most famous of the female institutes of England, the Somerville College, Oxford, founded in the early twentieth century, is named after her.

In spite of her outspoken feminism however (among other initiatives she signed the first petition in parliament for giving voting rights to women, a proposal of the philosopher and economist John Stuart Mill). At the same time she considered herself a woman who had been given exceptional opportunities, and her above mentioned quote on the alleged lack of originality and creativity of women for science remains an interesting contradiction.

In Italy Mary carried out a series of experiments on the action of the sun's rays on vegetable juices and sent her description and results to John Herschel who asked permission to publish parts of her letter.

Mary Somerville died on November 29, 1872 in Naples, at the age of 92, while she was working on an old mathematical article, a paper on quaternions, feeling regret for not having worked enough of this subject but as she said, with relentless persistence: *"Sometimes I find [mathematical problems] difficult, but my old obstinacy remains, for if I do not succeed today, I attack them again on the morrow."*

She is buried in the English cemetery there and much of her scientific library was donated to the recently founded Ledies' College at Hitchin and now Girton College for Women, in Cambridge.

After her death Martha, her daughter, published parts of her mother's memories and she was also immortalized in the designation of a lunar crater and an asteroid with her name.

25.3   As They Said of Her

Sir David Brewster, Principal of the University of St Andrews and inventor of the kaleidoscope, wrote in 1829: " *[...] certainly the most extraordinary woman in Europe – a mathematician of the very first rank with all the gentleness of a woman [...] She also a great natural philosopher and mineralogist."* Her friend Frances Power Cobbe thought Mary Somerville had to be buried in Westminster Abbey, but in the end her campaign was unsuccessful due to the opposition of George Airy, at the time Astronomer Royal and President of the Royal Society. According to her own words *"The Dean consented freely and with hearty approval to my proposition, and Mrs. Somerville's nephew, Sir William Fairfax, promised at once to defray all expenses. There was only one thing further needed, and that was the usual formal request from some public body or official persons to the Dean and Chapter of Westminster."* Such 'public body' had been identified in the person of the Astronomer Royal which, however, *"[...] refused to do it on the ground that he had never read Mrs. Somerville's books!"* The self-serving character of such an excuse is highlighted immediately after with the following, biting words: *"Whether he had read one in which she took the opposite side from his in the sharp and angry Adams-Le Verrier controversy, it is not for me to say."* Mary Somerville can be considered the last of the great amateur scientists and at the same time a lucky woman of science because, as Charles Lyell wrote to his future wife, Mary Corner, in 1831: *"If our friend Mary Somerville had been married to Laplace or some other mathematician, we would never have heard of her work. They would have incorporated into that of her husband and spread as his."*

Conclusions

This book contains biographies of female astronomers of the past, some of which are not so well known. Since the *trait-d'union* of these biographies was given by the "forgotten sister" which ideally Caroline Herschel mentions in the poem dedicated to her by Siv Cedering They intentionally stop between the eighteenth and the nineteenth centuries. The poem recalls just five names, but as we have seen there were quite a lot more spread everywhere in the world. Anyway, from antiquity to Caroline's contemporaries, our count stops at a few more than 20 names, so a question naturally arises: why are so few women of the past remembered in the history of astronomy before the famous astronomer Caroline? And why, once again before her, no female figures bear an importance at least comparable with that of their male colleagues?

Indeed, the list could be longer, but we included only the most significant names, or those for which there was an adequate amount of information, which brings us to the continuation of our reasoning. Actually, with the exception of some characters in the antiquity, when in any case the background was quite different from that of more recent times, the figures mentioned in the book were a kind of privileged ones. In fact women who, in past centuries, had the freedom and the ability to deal with science either received an unconventional education, compared to the canons of the time, or they had been introduced directly by some family members already well placed in the field. In this way their activity could cover various aspects of astronomy, such as the direct observation of stars, the calculation and data analysis or the writing of authoritative texts, but obviously they could hardly take a leading role in the field.

© Springer International Publishing Switzerland 2016
G. Bernardi, *The Unforgotten Sisters*, Springer Praxis Books in Popular Astronomy,
DOI 10.1007/978-3-319-26127-0

Distribution for the book sample of the main activity field

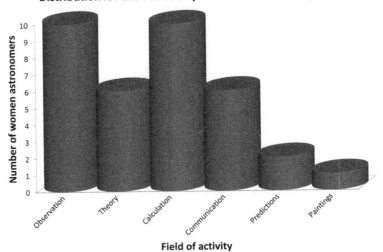

As shown by this plot which considers only the sample used for this book, the interests of the women astronomers of the past spanned all the fields of this subject, from observations to theoretical development

Moreover, they have been looked upon for a long time with suspicion and as exceptions, because the production of knowledge was reserved to men. Their names, therefore, have often been obliterated and their work incorporated into that of their male counterparts, as for example happened to "Madame G.," of whom we know only that she worked in anonymity with the French mathematician and astronomer Alexis Clairaut in the eighteenth century. In other cases they themselves signed with masculine names to lend credibility to their work, such as Antoine-August Le Blanc, the pseudonym that the French mathematician Sophie Germain used to correspond with her colleagues, the Italian Joseph-Louis Lagrange or the German Carl Friedrich Gauss.

Portrait of Sophie Germain at the age of 14

The twentieth century sees women entering into the scientific world and enjoying formal admission to universities both as students and as teachers, but still with a lack of equal career opportunities. Many graduates, in fact, had to be satisfied with the role of assistant or of high school teacher, and the female share among the attendants of scientific courses remained a tiny minority. In other circumstances, especially during war periods, the role of women was seen as a "buffer" in the labor market. They were engaged in tasks that were traditionally considered to be "masculine." Men were called to arms but, at the end of the conflicts and return of the "legitimate" worker, female workers usually had to step back.

An actual example can be found in Italy. Father Giovanni Boccardi, from 1903 to 1923, was Director of the Royal Astronomical Observatory of Turin, and in writing "*The new staff regulation of the Astronomical Observatories*" he expressed his doubts about the women's presence in Academia: "*[...] would it be prudent to put as an Assistant Professor, for example of descriptive Geometry and corresponding drawing, a young girl in the midst of 180 students? [...]*" this of course in case a seat had been won by a woman.

Under his direction, he probably had to face this problem, since the official records report the names of eight women initially enrolled as "Voluntary Assistant", most of whom came from the Science Faculty with a degree.

The various Astronomical Observatory Annual Reports quote their names and positions: Dr. **Luisa Viriglio**, graduated in mathematics and collaborator from 1904 to 1906, Dr. **Ernesta Fasciotti**, collaborator from 1905 to 1906. Dr. **Giovanna Greggi** enjoyed a more articulate career; collaborator from 1911 to 1912; in 1913 she was Second Assistant, while in 1927 she became Professor of mathematics and physics at the Technical Institute of Mondovì and wrote a mathematical work which was published in the 1911–12 proceedings of the *R. Istituto Veneto di Scienze, Lettere ed Arti* (Royal Venetian Institute of Science, Humanities and Arts). The list continues with the names of Dr. **Teresa Castelli**, collaborator from 1914 to 1918, Dr. **Tiziana Teresilla Comi**, graduated in mathematics with a thesis on the "Apparent curvature of planetary orbits" and collaborators for four years from 1914. During this period, in 1917, she published the *Ephemerides of the Sun and the Moon* and she was a member of the *Società Urania* (Urania Society) where she gave acclaimed presentations at conferences of astronomical topics. Dr. **Jeanette Mongini**, collaborator from 1919 to 1920, resigned from the post of Assistant on 25 November 1920. **Corinna Gualfredo**, started as a collaborator in 1916 by August 1916 and was appointed *Assistente di ruolo* (Staff Assistant) a role she kept until 1921. In the archives of the observatory one can also find a work of hers on the determination of the constants of the Meridian circle. The preface written by Director Boccardi is singular because he talks about her using the third masculine pronoun. Corinna Gualfredo joined the Editorial Board of the journal *Urania* and about her Boccardi wrote, once again: "*Mr. Gualfredo, who from 1915 onward had largely contributed to the observations, in particularly in 'I° verticale'* [i.e. the vertical circles normal to the meridian] *withdrew from observations at the end of September. Thus the Observatory is missing and effective, intelligent and unpaid aid.*" A quite clarifying statement. We have then **Carla Greggi**, "technical helper" from 1912 until 1920 who in a work by Corinna Gualfredo is defined as "calculator." This was the name that at that time defined people, usually women, dealing with boring astronomical calculations by hand, later replaced by actual mechanical or electromechanical calculators and finally by computers. The records contain also the name of **Lina Graneris**, who wrote a work on the "*Perturbation of Comet Pons-Winnecke in opposition of 1921 and 1927*", but afterwards the female presence at the observatory seems to be interrupted for about 15 years.

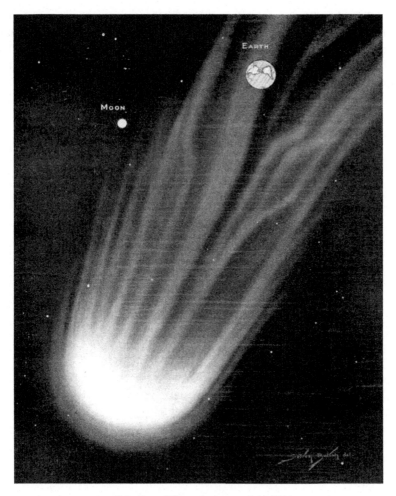

Contemporary artist concept of the Pons-Winnecke comet, July 1921

The next one in fact is Dr. **Francesca Demichelis**, who owned a degree in mathematics and physics and, coming from the Physics laboratory of the Polytechnic University of Turin, collaborated from 1942 until 1946 doing microphotometric diagrams used to study the traces of variable stars. Dr. **Ernesta Tedeschi**, collaborator from 1 December 1943 to 30 November 1946, graduated in 1942–1943 with a thesis on the *"Study of bright variations of AK Herculis's star and their interpretation in the hypothesis of his duplicity"*, and was then replaced by Dr. **Enrichetta Lagutaine** who graduated in 1945–1946 with a thesis on the *"Critical examination of the theory by Russel for the determination of orbits of photometric binary systems."* She was hired first in December 1, 1946 for a single year at the Observatory of Turin as "appointed for studies and research". Finally, there appears the name of **Teresina Tamburini**, whose official job was reorganization of the library after the occupation, and was employed at the Observatory in 1966. Today the situation is very different. The female staff is nearly 50 % at all levels, and years ago the Directorship was given to a woman, but the memory of these not too distant "sisters" that preceded them should not fall into oblivion.

Bibliography

Books and Articles

Ken Alder, *The Measure of All Things: the seven-year odyssey and hidden error that transformed the world*, Free Press, New York, 2002

Margaret Alic, *L'eredità di Ipazia. Donne nella storia delle scienze dall'antichità all'Ottocento*, Editori Riuniti, 1989

Alicchio, R. Pezzoli, C., (a cura di) *Donne di scienza : esperienze e riflessioni*, Torino, Rosenberg & Sellier, 1988

Alma Mater Studiorum, *La presenza femminile dal XVIII al XX secolo*, CLUEB Bologna, 1988

Emilio Ambrisi, Franco Eugeni, *La donna e il mondo della scienza - dall'antichità all'epoca dei lumi*, Atti I Simposio Internazionale Elbano di Studi Filosofici, Centro Sociologico Italiano, Roma 1995

Annuari del Regio Osservatorio Astronomico di Pino Torinese (1902–1945)

Barbara Bennett Peterson, He Hong Fei, Guangyu Zhang, *Notable Women Of China: Shang Dynasty To The Early Twentieth Century*, Routledge, 2000

Gemma Beretta, *Ipazia d'Alessandria*, Editori Riuniti, 1992

Giovanni Boccardi, Il nuovo regolamento per personale degli Osservatori Astronomici-1912, Saggi di astronomia popolare

Mary Brück, *Women in Early British and Irish Astronom. Stars and Satellites*, Springer, 2009

Paul J. Campbell, Louise S. Grinstein, *Women of Mathematics*, A Bibliographic Soucebook, 1987

Siv Cedering, *Letters from the Floating World: New and Selected Poems*, University of Pittsburgh Press, 1984

Allan Chapman, *Mary Somerville and the world of Science*, Springer 2014

John Robert Christianson, *On Tychos island: Tycho Brahe and his assistants, 1570–1601*, Cambridge university press

Christianson, John Robert (2000). *On Tycho's Island: Tycho Brahe and his assistants, 1570–1601*. Cambridge University Press. ISBN 0-521-65081-X

Agnes Mary Clerke, *The Herschel and Modern Astronomy*, Cambridge University Press, 2010

Agnes Mary Clerke, Caroline Lucretia Herschel, Dictionary of National Biography XXVI, London, 1891, 260–263

Sylvie Coyaud, La scienza: una passione femminile. Scienziate grandi e grandissime nella storia e nell'attualità. In Cleis Franca, Varini Ferrari Osvalda (a cura di), (2001) *Pensare un mondo con le donne. Saperi femminili nella scienza, nella società e nella letteratura*. Atti del corso di

© Springer International Publishing Switzerland 2016
G. Bernardi, *The Unforgotten Sisters*, Springer Praxis Books in Popular Astronomy,
DOI 10.1007/978-3-319-26127-0

formazione sulla presenza femminile nella storia e nella cultura del XX secolo (anni 1996–1999)

Giorgio Dragoni, Silvio Bergia, Giovanni Gottardi, *Dizionario Biografico degli Scienziati e dei Tecnici*, Zanichelli, 1999

European Commission, *Women in Science*, Luxembourg: Office for Official Publications of the European Communities, 2009, ISBN 978-92-79-11486-1

Garzya Antonio, *Opere di Sinesio di Cirene: epistole, operette, inni,* Turin: Unione Tipografico-Editrice Torinese, 1989

Corinna Gualfredo, *Il piccolo cerchio meridiano dell'Osservatorio di Pino Torinese: determinazioni delle costanti*, Tip. Artigianelli, 1916

Segura Graiño, C (1998). *Diccionario de mujeres en la historia*. Madrid: Espasa-Calpe

Margherita Hack, I contributi delle donne alla scienza: ieri e oggi, *Pristem*, Università Bocconi

Mrs. John Herschel [Mary Herschel]. *Memoirs and Correspondence of Caroline Herschel*. John Murray, Albemarle Street, London, 1876.

Hildegard von Bingen, *Scivias,* trans. by Columba Hart and Jane Bishop with an Introduction by Barbara J. Newman, and Preface by Caroline Walker Bynum (New York: Paulist Press, 1990) 60–61

Thomas Hockey, Katherine Bracher, Marvin Bolt, Virginia Trimble, JoAnn Palmeri, Richard Jarrell, Jordan D. Marché, F. Jamil Ragep, *Biographical Encyclopedia of Astronomers*, Springer 2007

Hoskin Micheal, *Caroline Herschel's Autobiographies,* Science History Publications Ltd, Cambridge, 2003

Michael Hoskin, 2006. Caroline Herschel's Catalogue of Nebulae. *Journal for the History of Astronomy*, Vol. 37, Part 3, No. 128, pp. 251–253 [ADS: 2006JHA....37..251H]

Michael Hoskin, 2005. Caroline Herschel as Observer. *Journal for the History of Astronomy*, Vol. 36, Part 4, No. 125, pp. 373–406 (November 2005). [ADS: 2005JHA....36..373H]

Michael Hoskin, Caroline Herschel: assistant astronomer or astronomical assistant?, *History of Science* (2002): 425–444

Jerome Lalande, *Bibliographie Astronomique: avec l'histoire de l 'astronomie depuis 1781 jusqu'a 1802*, Paris, 1803

Jerome Lalande, *Astronomie des dames,* Paris, 1790

Leigh Ann Whaley, *Women's History as Scientists: A Guide to the Debates*, ABC-CLIO, 2003

Erika Luciano, Clara Silvia Roero, *Numeri, Atomi e Alambicchi-Donne e Scienza in Piemonte dal 1840 al 1960 Parte I*, Centro Studi e Documentazione Pensiero Femminile 2008

Arthur I. Moller, *L'equazione dell'anima*, Corriere della Sera, 2009

L S Multhauf, Biography in *Dictionary of Scientific Biography*, New York , 1970–1990.

Kathryn A. Neeley, *Mary Somerville: Science, Illumination and the Female Mind*, Cambridge University Press, 2001

Mariolyn Ogilvie, Joy Harvey, *The Biographical Dictionary of Women in Science*, Routledge, 2000

Mariolyn Ogilvie, Joy Harvey, *The Biographical Dictionary of Women in Science*, Routledge, 2003

Bailey Ogivie, *Women in Science,* The MIT Press Cambridge, 1990

Ogilvie, Marilyn Bailey *Women in Science: antiquity through the nineteenth century.* Boston: Massechussets Institute of Technology, 1986.

Franco Pastrone, Il progetto WIND come spunto per un'analisi dei percorsi universitari nelle Facoltà scientifiche torinesi, *Associazione Subalpina Mathesis*

William W. Payne; H.C. Wilson (1898). "Maria Clara Muller". *Popular Astronomy Vol. 6.* Minnesota: Goodsell Observatory of Charleton College

Jean Pierre Poirier, *Histoire des fammes de science en France,* Pygmalion Gérard Watelet, 2002

Terry Pratt, David McCallam, David Williams, *The Enterprise of Enlightenment: A Tribute to David Williams from His Friends,* Peter Lang, 2004

F J Ring, John Herschel and his heritage, in D G King-Hele (ed.), *John Herschel 1792–1871: A bicentennial commemoration* (London, 1992), 3–16

Vera Rubin, *Bright Galaxies, Dark Matters*, Springer Science & Business Media 1997

Sara Sesti, Liliana Moro, *Donne di Scienza*, Centro ELEUSI-PRISTEM Università Bocconi, 2002

Sara Sesti, Liliana Moro, *Scienziate nel tempo. 70 biografie*, edizioni LUD, Milano 2010

Madigan, Shawn. *Mystics, Visionaries and Prophets: A Historical Anthology of Women's Spiritual Writings*, Minnesota: Augsburg Fortress, 1998

Londa Schiebinger, *The Mind Has No Sex? Women in the Origins of Modern Science*, Cambridge: Harvard University Press, 1991

Londa Schiebinger, Maria Winkelmann at the Berlin Academy: A Turning Point for Women in Science, *Isis* 78 (1987): 174–200.

Londa Schiebinger, Women in Science: historical perspective, In C. Megan Urry, Laura Danly, Lisa E. Sherbert and Shireen Gonzaga (Eds.) *Women at Work: A Meeting on the Status of Women in Astronomy*. Proceedings of the conference held at the Space Telescope Science Institute, Baltimore, September 8–9, 1992

Edward Singleton Holden, *Sir William Herschel: His Life and Works*, New York 1881

Mary Somerville, *Personal Recollections, from Early Life to Old Age*, London 1874

Vignoles, Alphonse des. Eloge de Madame Kirch à l'occasion de laquelle on parle de quelques autres femmes & d'un paison astronomes. *Bibliothèque germanique* 3 (1721): 115–183.

Lémonon Waxin, *Domesticity in the Making of Modern Science*, Palgrave, 2015

Margaret Wertheim, *I pantaloni di Pitagora*. Dio, le donne e la Matematica, Torino, Instar Libri, 1996

Yung-Chung Kim, Women of Korea, *A History from Ancient Times to 1945*, Seoul: EWHA Women's University Press, 1997

Web

http://www.stsci.edu/institute/smo/visitor-programs/caroline-herschel/poem

http://www-history.mcs.st-and.ac.uk/

http://www.sheisanastronomer.org/index.php/history

http://scienzaa2voci.unibo.it/biografie

http://users.clas.ufl.edu/ufhatch/pages/03-Sci-Rev/SCI-REV-Home/sr-women/05-SR-WOMEN-CUNITZ-PAGE-bio.html

http://www.brainpickings.org/2014/08/11/johannes-hevelius-catalog-of-stars/

http://pionnieres.revolublog.com/jeanne-dumee-a44539607

http://www.naa.net/ain/personen/show.asp?ID=39

http://www.bo.astro.it/dip/Museum/english/car_67.html

http://www.distinguishedwomen.com/

http://biography.yourdictionary.com/maria-winckelmann-kirch

http://www.epigenesys.eu/en/science-and-you/women-in-science

http://www.theworldofchinese.com/2014/10/badass-ladies-of-chinese-history-wang-zhenyi/

http://www.agnesscott.edu/lriddle/women/alpha.htm

http://orlando.cambridge.org/public/svPeople?person_id=bryama

http://herschelmuseum.org.uk/history/star-objects/

http://turnbull.mcs.st-and.ac.uk/history/

http://matematica-old.unibocconi.it/donne/hack.htm#up